TOM WORLEY 9E

Modern Methods of Plant Analysis

New Series Volume 5

Editors
H. F. Linskens, Erlangen
J. F. Jackson, Adelaide

Volumes Already Published in this Series:

Volume 1: Cell Components
1985, ISBN 3-540-15822-7

Volume 2: Nuclear Magnetic Resonance
1986, ISBN 3-540-15910-X

Volume 3: Gas Chromatography/
Mass Spectrometry
1986, ISBN 3-540-15911-8

Volume 4: Immunology in Plant Sciences
1986, ISBN 3-540-16842-7

Volume 5: High Performance Liquid Chromatography
in Plant Sciences
1987, ISBN 3-540-17243-2

High Performance Liquid Chromatography in Plant Sciences

Edited by
H.F. Linskens and J.F. Jackson

Contributors

G.W.M. Barendse Ch.-M. Chen R.A. Creelman
A. Crozier P.J. Davies A. Ernstsen A.W. Galston
R.D. Hartley J.P.F.G. Helsper L.E. Hood S.B.H. Kent
J.M. Robertson G. Sandberg L.A. Smith M.A. Smith
R.N. Strange R. Sütfeld B. Sundberg P. Tempst
G.A. Thompson, Jr. A.F. Tiburcio J.A.D. Zeevaart

With 74 Figures

Springer-Verlag
Berlin Heidelberg New York
London Paris Tokyo

Professor Dr. Hans-Ferdinand Linskens
Goldberglein 7
D-8520 Erlangen

Professor Dr. John F. Jackson
Department of Biochemistry
Waite Agricultural Research Institute
University of Adelaide
Glen Osmond, S.A. 5064
Australia

ISBN 3-540-17243-2 Springer-Verlag Berlin Heidelberg New York
ISBN 0-387-17243-2 Springer-Verlag New York Berlin Heidelberg

Library of Congress Cataloging-in-Publication Data. High performance liquid chromatography in plant sciences. (Modern methods of plant analysis; new ser., v. 5) Includes bibliographies and index. 1. Plants – Analysis. 2. High performance liquid chromatography. I. Linskens, H.F. (Hans F.), 1921–. II. Jackson, J.F. (John F.), 1935–. III. Barendse, G.W.M. (Gerard W.M.) IV. Series. QK865.H58 1987 581′.01′5430894 87-4303
ISBN 0-387-17243-2 (U.S.)

Typesetting, printing and bookbinding: Brühlsche Universitätsdruckerei, Giessen
2131/3130-543210

Introduction

Modern Methods of Plant Analysis

When the handbook *Modern Methods of Plant Analysis* was first introduced in 1954 the considerations were:

1. the dependence of scientific progress in biology on the improvement of existing and the introduction of new methods;
2. the difficulty in finding many new analytical methods in specialized journals which are normally not accessible to experimental plant biologists;
3. the fact that in the methods sections of papers the description of methods is frequently so compact, or even sometimes so incomplete that it is difficult to reproduce experiments.

These considerations still stand today.

The series was highly successful, seven volumes appearing between 1956 and 1964. Since there is still today a demand for the old series, the publisher has decided to resume publication of *Modern Methods of Plant Analysis*. It is hoped that the New Series will be just as acceptable to those working in plant sciences and related fields as the early volumes undoubtedly were. It is difficult to single out the major reasons for success of any publication, but we believe that the methods published in the first series were up-to-date at the time and presented in a way that made description, as applied to plant material, complete in itself with little need to consult other publications.

Contributing authors have attempted to follow these guidelines in this New Series of volumes.

Editorial

The earlier series *Modern Methods of Plant Analysis* was initiated by Michel V. Tracey, at that time in Rothamsted, later in Sydney, and by the late Karl Paech (1910–1955), at that time at Tübingen. The New Series will be edited by Paech's successor H.F. Linskens (Nijmegen, The Netherlands) and John F. Jackson (Adelaide, South Australia). As were the earlier editors, we are convinced "that there is a real need for a collection of reliable up-to-date methods for plant analysis in large areas of applied biology ranging from agriculture and horticultural experiment stations to pharmaceutical and technical institutes concerned with raw material of plant origin". The recent developments in the fields of plant biotechnology and genetic engineering make it even more important for workers in the plant sciences to become acquainted with the more sophisticated methods, which sometimes come from biochemistry and biophysics, but which also have been developed in commercial firms, space science laboratories, non-university research institutes, and medical establishments.

Concept of the New Series

Many methods described in the biochemical, biophysical, and medical literature cannot be applied directly to plant material because of the special cell structure, surrounded by a tough cell wall, and the general lack of knowledge of the specific behavior of plant raw material during extraction procedures. Therefore all authors of this New Series have been chosen because of their special experience with handling plant material, resulting in the adaptation of methods to problems of plant metabolism. Nevertheless, each particular material from a plant species may require some modification of described methods and usual techniques. The methods are described critically, with hints as to their limitations. In general it will be possible to adapt the methods described to the specific needs of the users of this series, but nevertheless references have been made to the original papers and authors. While the editors have worked to plan in this New Series and made efforts to ensure that the aims and general layout of the contributions are within the general guidelines indicated above, we have tried not to interfere too much with the personal style of each author.

Volume Five – High Performance Liquid Chromatography in Plant Sciences

The New Series in Modern Methods of Plant Analysis was begun in 1985 with a volume on Cell Components, and quickly followed by three further volumes. These dealt with the powerful analytical techniques of Nuclear Magnetic Resonance (Vol. 2), Gas Chromatography/Mass Spectrometry (Vol. 3), and a volume on Immunology in Plant Sciences (Vol. 4). The latter presents a series of chapters on the application of the antigen-antibody interaction to the detection of a large range of plant substances. These very specific immuno-detection techniques are so often used in conjunction with, or in addition to, both GC/MS and High Performance Liquid Chromatography (HPLC), that it is appropriate that the volume before it dealt with GC/MS and the volume after (i.e., the present Vol. 5) with HPLC.

Thus a very complete chapter on gibberellins through HPLC is given by Barendse in this Volume, a chapter which compliments that by Hedden in Vol. 3 (GC/MS) and by Weiler in Vol. 4 (Immunology). Similarly in this Volume, Chen gives methods for the characterization of cytokinins by HPLC; these growth factors are also dealt with in the volumes on GC/MS and on Immunology. Volumes 3, 4, and 5 in the New Series are therefore good companion volumes for those wanting to determine and characterize fully the plant hormones and other growth factors.

In addition to the gibberillins and cytokinins, this Volume includes chapters on indole acetic acid, abscisic acid, and the polyamines. Further chapters describe the use of HPLC for the determination of cell wall aromatic compounds, for ascorbic acid, and for thiophene derivatives.

Phytoalexins are also the subject of a chapter in this book, as well as the purification and analysis of polypeptides by HPLC. No volume on plant products would be complete without a treatment of the alkaloids; Galston and Tiburcio therefore give methods for plant alkaloids using HPLC. Plant lipid analysis is often carried out by means of thin layer chromatography (TLC) or gas liquid chromatography (GLC), but Thompson demonstrates here that HPLC can be used for this purpose also.

HPLC is now a firmly established separation technique which can be used in a large variety of ways in a laboratory dealing with plant products, both for preparative and analytical purposes. Recent developments have come in the commercial availibity of a large variety of HPLC columns for specialist separations such as for organic acids, carbohydrates and proteins, and the development of guard columns to protect other columns in the high pressure system. Different solvents are now being used, the detectors in the main being spectrophotometric and fluorometric based.

The present volume shows the utility of HPLC in detecting and characterizing a large spectrum of plant products, and should be used in conjugation with its companion Volumes 3 and 4 dealing with GC/MS and immunological techniques for a complete and definitive detection system for plant hormones and other plant products.

The editors believe that this fifth Volume will find immediate use for undergraduate laboratory classes, as well as for research laboratories. We hope that this collection of chapters by world experts on HPLC will further encourage the spread of this technique among scientists, students, and industrial analysts working with plant materials.

Acknowledgements. The editors express their thanks to all contributors for their efforts in keeping to production schedules, and to Dr. Dieter Czeschlik, Ms. K. Gödel and Ms. I. Samide of Springer-Verlag for their co-operation with this and other volumes in Modern Methods of Plant Analysis. The constant help of José Broekmans is gratefully acknowledged.

Nijmegen and Adelaide, April 1987

H. F. Linskens
J. F. Jackson

Contents

HPLC of Thiophenes for Phytochemical and Biochemical Research
R. Sütfeld (With 6 Figures)

High Performance Liquid Chromatography of Ascorbic Acid
J. P. F. G. HELSPER

High Performance Liquid Chromatography of Phytoalexins
R. N. STRANGE (With 13 Figures)

Practical High Performance Liquid Chromatography of Proteins and Peptides

P. TEMPST, L. E. HOOD, and S. B. H. KENT (With 7 Figures)

Monitoring Polyamines in Plant Tissues by High Performance Liquid Chromatography
M. A. SMITH and P. J. DAVIES (With 7 Figures)

Analysis of Alkaloids in Tobacco Callus by HPLC
A. F. TIBURCIO and A. W. GALSTON (With 3 Figures)

List of Contributors

BARENDSE, GERARD W. M., Department of Botany, University of Nijmegen, Toernooiveld, 6525 ED Nijmegen, The Netherlands

CHEN, CHONG-MAW, Biomedical Research Institute, University of Wisconsin-Parkside, Kenosha, WI 53141, USA

CREELMAN, ROBERT A., Department of Biochemistry and Biophysics, Texas A+M University, College Station, TX 77843-2128, USA

CROZIER, ALAN, Department of Botany, The University, Glasgow G12 8QQ, Scotland, United Kingdom

DAVIES, PETER J., Section of Plant Biology, Cornell University, Ithaca, NY 14853, USA

ERNSTSEN, ARILD, Department of Forest Genetics and Plant Physiology, The Swedish University of Agricultural Sciences, 90183 Umeå, Sweden

GALSTON, ARTHUR W., Department of Biology, Yale University, P.O. Box 6666, New Haven, CT 06511, USA

HARTLEY, ROY D., Institute for Grassland and Animal Production, Hurley, Maidenhead, Berkshire SL6 5LR, United Kingdom

HELSPER, JOHANNES, P. F. G., Research Institute, ITAL, P.O. Box 48, 6700 AA Wageningen, The Netherlands

HOOD, LEROY E., Division of Biology 156-29, California Institute of Technology, Pasadena, CA 91125, USA

KENT, STEPHEN B. H., Division of Biology 147-75, California Institute of Biology, Pasadena, CA 91125, USA

ROBERTSON, J. MASON, Plant Physiology Research Group, Biology Department, University of Calgary, Calgary, Alberta, Canada T2N 1N4

SANDBERG, GÖRAN, Department of Forest Genetics and Plant Physiology, The Swedish University of Agricultural Sciences, 90183 Umeå, Sweden

SMITH, LAURENCE A., Procter and Gamble Co., Ivorydale Technology Center, Room 2W-66, S 299 Spring Grove Ave., Cincinnati, OH 45217, USA

SMITH, MARY A., MSU-DOE Plant Research Laboratory, Michigan State University, East Lansing, MI 48824, USA

STRANGE, RICHARD N., Department of Botany and Microbiology, University College London, Gower Street, London WC1E 6BT, United Kingdom

SÜTFELD, RAINER, Botanisches Institut der Westfälischen Wilhelms-Universität, Schloßgarten 3, 4400 Münster, FRG

SUNDBERG, BJÖRN, Department of Forest Genetics and Plant Physiology, The Swedish University of Agricultural Sciences, 90183 Umeå, Sweden

TEMPST, PAUL, Howard Hughes Medical Institute, Department of Genetics, Harvard Medical School, 25 Shattuck Street, Boston, MA 02115, USA

THOMPSON, Jr. GUY A., Department of Botany, University of Texas, Austin, TX 78713, USA

TIBURCIO, ANTONIO F., Laboratori de Fisiologia Vegetal, Facultat de Farmacia, Universitat de Barcelona, Barcelona 08028, Spain

ZEEVAART, JAN A. D., MSU-DOE Plant Research Laboratory, Michigan State University, East Lansing, MI 48824, USA

High Performance Liquid Chromatography of Gibberellins

G. W. M. Barendse

1 Introduction

The ability to identify and accurately measure plant hormones has become increasingly important in studies on the physiology and biochemistry of plant hormones. The analysis of plant hormones such as the gibberellins in plant extracts poses many technical problems, primarily because only minute amounts are present. Their concentration in fresh material rarely exceeds 1 part in 10^6, but is often as low as 1 part in 10^{10} and therefore they require analytical procedures which are able to measure ng or rather pg quantities of hormone present in extracts. In addition, plant extracts contain many more low molecular compounds than animal tissue, and these are difficult to separate from the low molecular plant hormones. Thus plant hormone analysis, and that of gibberellins in particular, is not yet a routine procedure, and at the present time not one particular method can be singled out as the best way to analyze plant hormones.

The criteria to be applied in the evaluation of a particular analytical method have been reviewed by Brenner (1981) and Reeve and Crozier (1980).

The gibberellins pose special problems, since at the time of this report 68 different gibberellins (Fig. 1), 7 GA-glucosyl ethers, 7 GA-glucosyl esters plus related compounds such as gibberethione and antheridiogen are known to occur. All gibberellins and related compounds have the ent-gibberellane skeleton (top Fig. 1). Extracts of a single plant species may contain a variety of gibberellins including free gibberellins, gibberellin glucosides, and gibberellin glucosyl esters, e.g., seeds of *Phaseolus vulgaris* (Hiraga et al. 1974). Therefore, purification procedures of gibberellins are not so simple. One must often choose a combination of different analytical procedures adapted to the particular material under study.

With gibberellins, combined GC-MS has become the physical method of choice for rigorous analysis of plant extracts (Binks et al. 1969). However, GC-MS requires highly purified extracts, which are not always attainable, or only at the expense of great losses during the prepurification steps. Secondly, GC-MS represents highly sophisticated instrumentation only available to a limited number of laboratories, which prevents it from becoming an instrument for routine analysis.

Recent developments in high performance liquid chromatography (HPLC) have shown that this technique is now well suited for plant hormone analysis in general, and for gibberellins in particular. The availability of a rapidly increasing array of column supports with different selectivities and high column efficiencies which allow rapid speeds of analysis at ambient temperatures all contribute to the

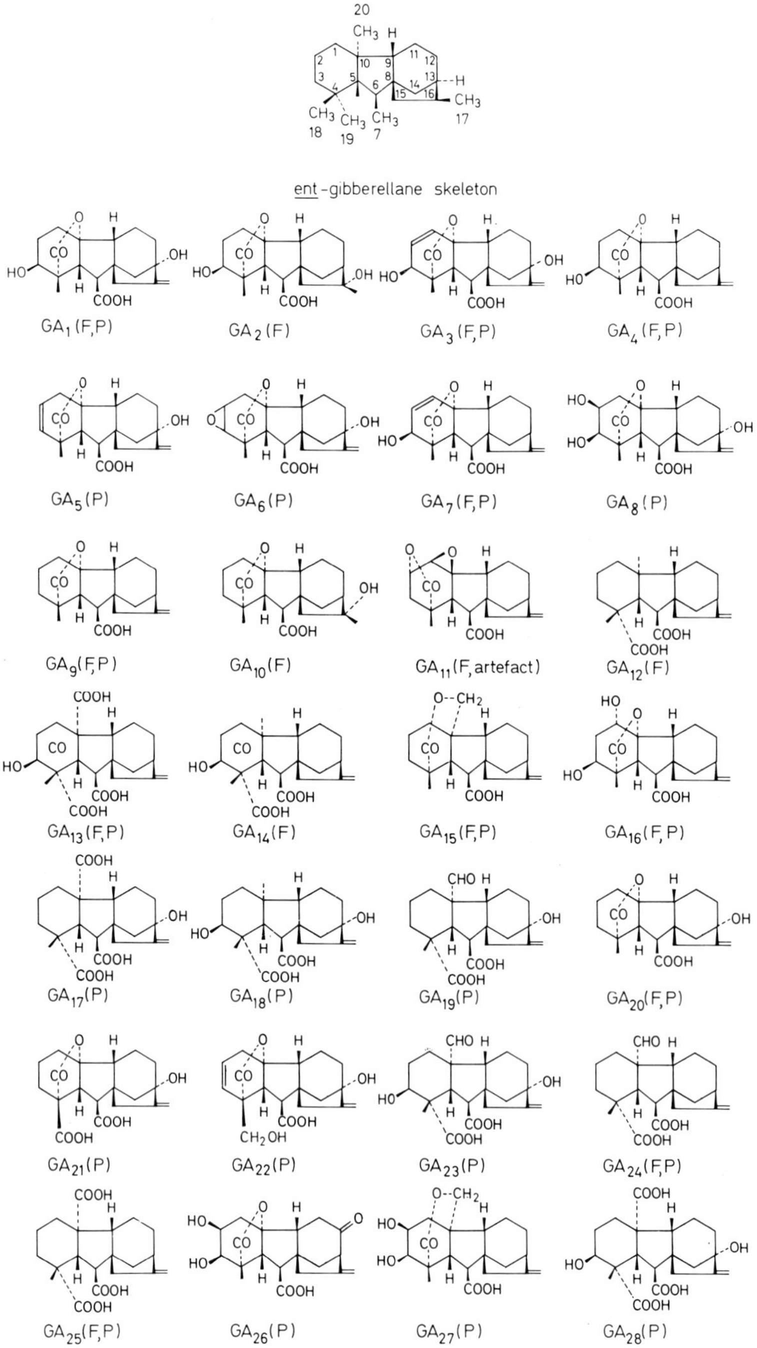

Fig. 1. Structure of free gibberellins (GA's), conjugated GA's and related compounds. *P* and *F* in parentheses denote origins; plant (*P*) and fungus (*F*). (Takahashi et al. 1985)

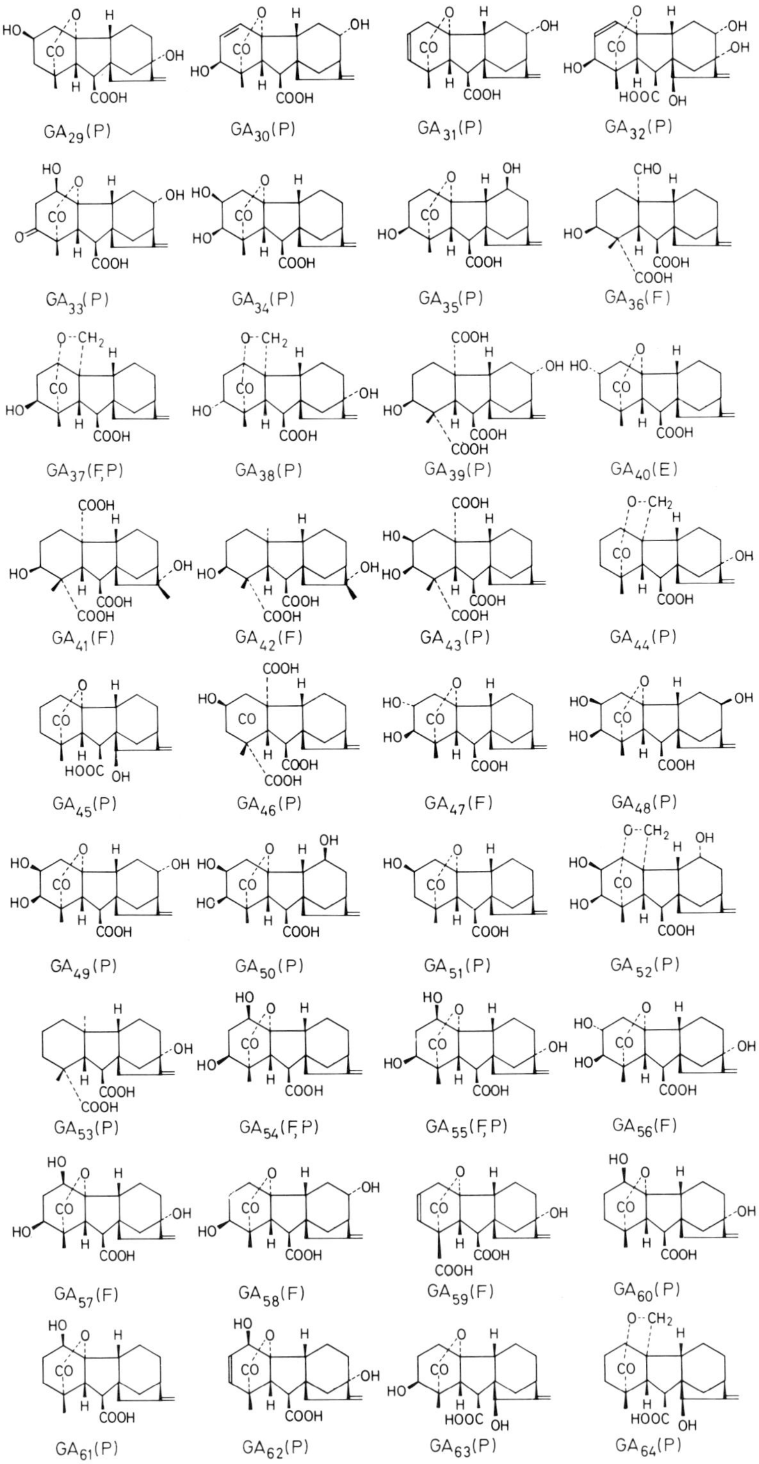
GA29(P)
GA30(P)
GA31(P)
GA32(P)
GA33(P)
GA34(P)
GA35(P)
GA36(F)
GA37(F,P)
GA38(P)
GA39(P)
GA40(E)
GA41(F)
GA42(F)
GA43(P)
GA44(P)
GA45(P)
GA46(P)
GA47(F)
GA48(P)
GA49(P)
GA50(P)
GA51(P)
GA52(P)
GA53(P)
GA54(F,P)
GA55(F,P)
GA56(F)
GA57(F)
GA58(F)
GA59(F)
GA60(P)
GA61(P)
GA62(P)
GA63(P)
GA64(P)

Fig. 1 (cont.)

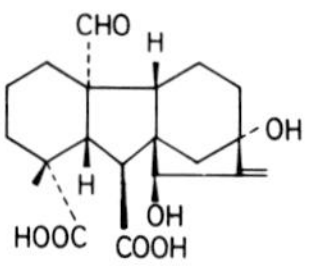
GA_{65} (P)

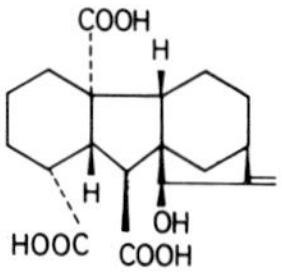
GA_{66} (P)

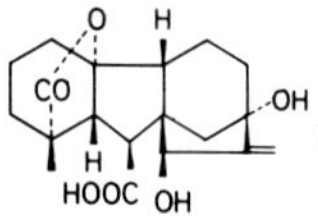
GA_{67} (P)

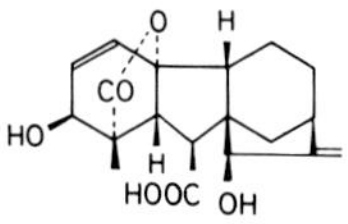
GA_{68} (P)

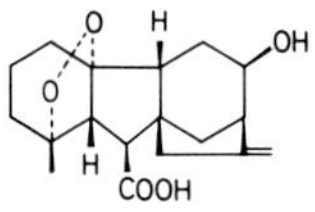
GA_{69} (P)

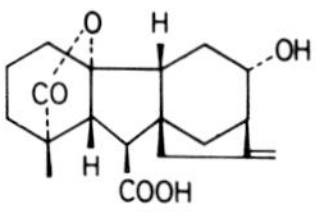
GA_{70} (P)

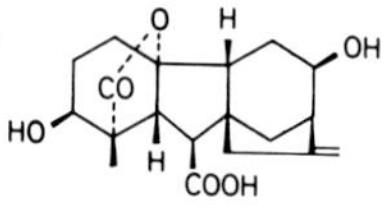
GA_{71} (P)

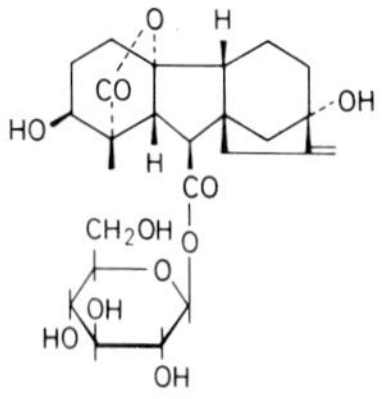
GA_1 Glucosyl ester

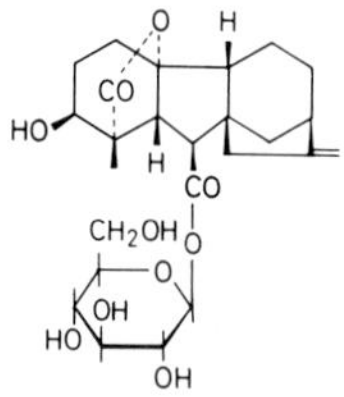
GA_4 Glucosyl ester

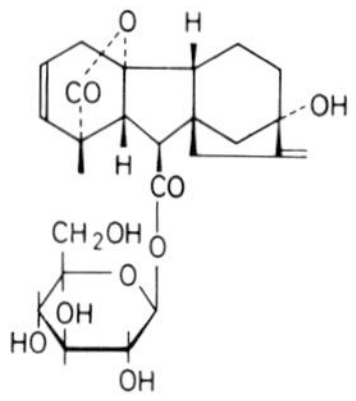
GA_5 Glucosyl ester

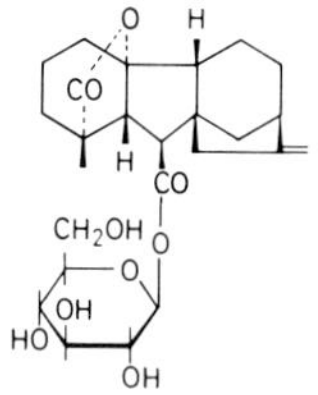
GA_9 Glucosyl ester

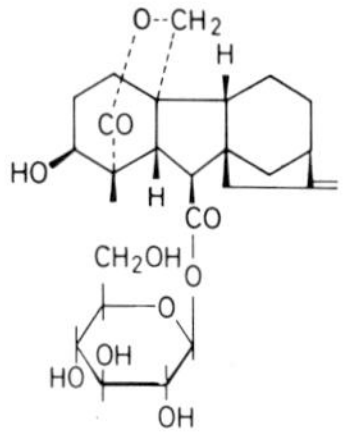
GA_{37} Glucosyl ester

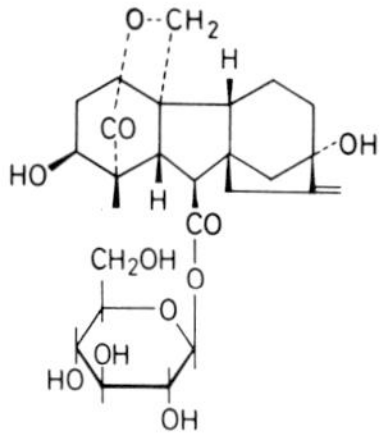
GA_{38} Glucosyl ester

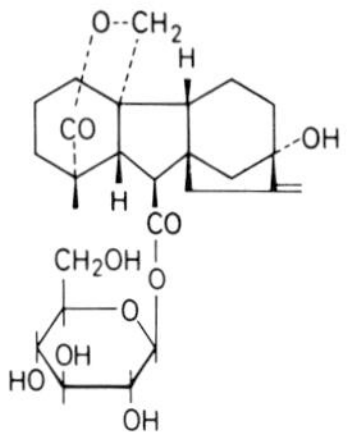
GA_{44} Glucosyl ester

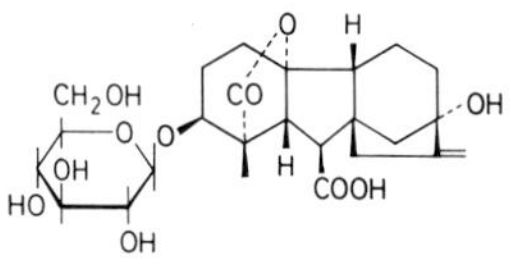
GA_1 Glucosyl ether

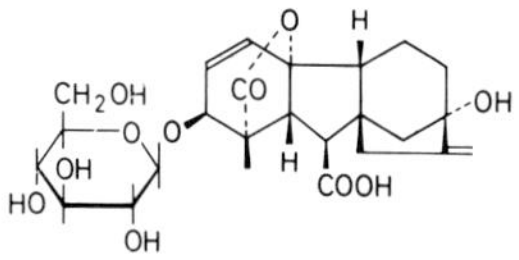
GA_3 Glucosyl ether

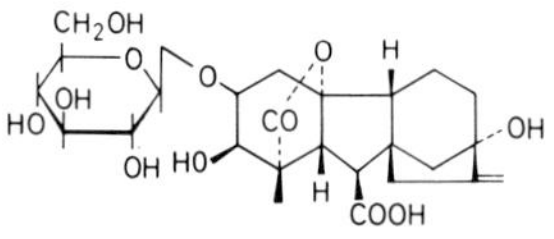
GA_8 Glucosyl ether

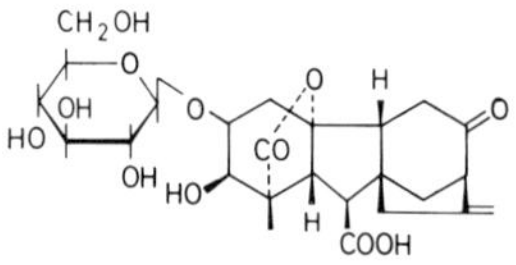
GA_{26} Glucosyl ether

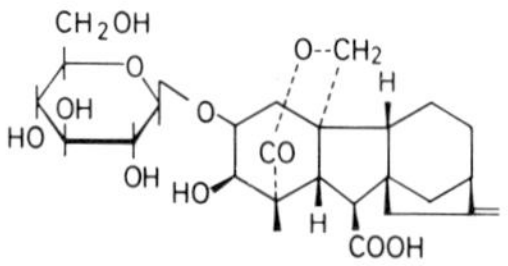
GA_{27} Glucosyl ether

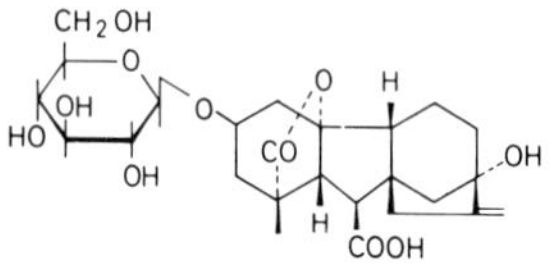
GA_{29} Glucosyl ether

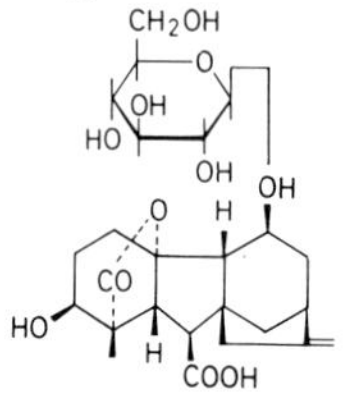
GA_{35} Glucosyl ether

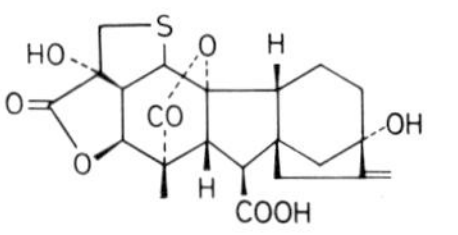
Gibberethione

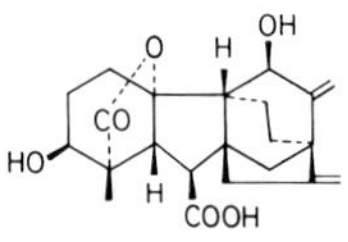
Antheridiogen (A_{An})

Fig. 1 (cont.)

overall effectiveness of HPLC (Knox 1979). HPLC can be employed both as preparative and analytical instrumentation which can be used in combination. In addition, several sensitive detector systems are now on the market for HPLC which, particularly when used in combination, have enhanced both sensitivity and selectivity of HPLC tremendously compared with the early low sensitivity as well as low selectivity of UV detector systems. Thus HPLC promises to become the cornerstone of gibberellin isolation and analysis, and is easily combined with other techniques as diverse as bioassay, immunoassay, and GC-MS (see vols. 3 and 4 of this series).

2 Extraction and Purification

Before HPLC can be employed, the GA's have to be isolated and purified. For GA's several isolation procedures have been reported, and each procedure has its own characteristics to make it most appropriate for the extraction and separation of certain GA's (Takahashi et al. 1985). Originally, relatively large amounts of GA's were required for structural studies and identification, which involved extremely large amounts of materials for extraction and elaborate purification steps. To detect the biological active fractions during the purification steps, several GA bioassays are used (Reeve and Crozier 1980).

Recent developments in analytical instrumentation with greatly enhanced sensitivity, resolution, and reproducibility, including the accumulation of spectroscopic data on GA's, have considerably diminished the amount of material required for structure determination, as well as for quantitative analysis.

In general, plant material is usually extracted by homogenization in methanol, or less frequently in 80% aqueous acetone, followed by filtration which is repeated three or four times. The organic solvent in the filtrate is removed in vacuo and the residual aqueous phase submitted to solvent partitioning. A representative scheme is presented in Fig. 2 (Takahashi et al. 1985), which yields an acidic ethyl acetate (AE), a neutral ethyl acetate (NE), a neutral n-butanol (NB) and an acidic n-butanol (AB) fraction. Most of the free gibberellins partition into the AE fraction, except for GA_9, GA_{12}, and GA_{25}, which sometimes appear in the NE fraction, while GA_{28} is not completely extracted with ethyl acetate, GA_{32} is almost nonextractable with ethyl acetate, but can be extracted with n-butanol into the NB fraction.

All the free gibberelins are unstable in acid. In addition, gibberellins with a C-1 $\approx$ C-2 double bond and a 3β-hydoxyl group such as GA_3 and those carrying a α-hydroxyl group plus a lactone linkage from C-4 to C-10 (γ-lactone) or to C-20 (δ-lactone) are unstable under basic conditions, which leads to isomerization of the 3β-hydroxyl group (Takahashi et al. 1985).

Under acidic, e.g., pH 2.5–3.0, or basic conditions, e.g., pH 8.0–8.5, GA's are usually stable for relatively short periods of time. Conjugated gibberellins show characteristics different from the ethyl acetate-soluble free GA's in the solvent partitioning procedure. Glucosyl-ether gibberellins are not extracted with ethyl acetate but can be partitioned with n-butanol into the AB fraction (Fig. 2). As yet

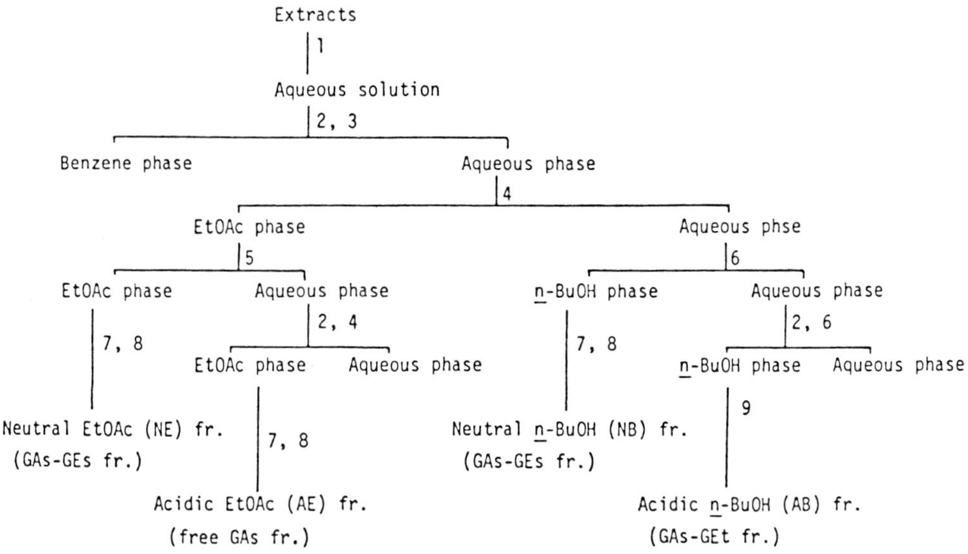

Fig. 2. General solvent fractionation procedure for gibberellins. *1* evaporation of organic solvent in vacuo; *2* pH adjustment to 2.5–3.0; *3* extraction with benzene (3 ×); *4* extraction with ethyl acetate (3 ×); *5* extraction with aqueous $NaHCO_3$; *6* extraction with n-butanol; *7* dried over anhydrous Na_2SO_4; *8* concentration in vacuo; *9* pH adjustment to 6.5 and concentration in vacuo. (Takahashi et al. 1986)

an adequate solvent fractionation procedure has not been established for the GA glucosyl esters (Takahashi et al. 1985). Some of the GA glucosyl ester can be partitioned with ethyl acetate, while the GA_1- and GA_{38} glucosyl esters are only partially extracted with this solvent. Complete extraction of these GA glucosyl esters can be obtained with n-butanol partitioning (Takahashi et al. 1985). Fractions containing GA glucosyl esters should not be stored under basic conditions, since their aglycones are then readily released. Certain gibberellin conjugates also migrate into ethyl acetate during partitioning; which proportion of the total conjugated GA's present is not known, since their partitioning behavior is as yet not well understood. Precautions have to be taken with regard to evaporating the ethyl acetate-soluble fractions to dryness, since this may cause hydrolysis of the GA conjugates. The fractions can be dried first with anhydrous sodium sulfate to reduce their water content to below 8%, which corresponds to the composition of the ethyl acetate/water aseotrope (Reeve and Crozier 1978) or by evaporating the solvent in vacuo to approximately a quarter of its original volume before adding a large excess of toluene and removing the aqueous ethyl acetate as the toluene/ethyl acetate/water ternary aseotrope.

However, in practice it appears difficult to dry the ethyl acetate thoroughly with anhydrous sodium sulfate so that both procedures can be performed. It is obvious from the foregoing that isolation and purification of gibberellins from plant materials is rather complex, and no single procedure can be outlined which will recover all known gibberellins and their conjugates under any conditions or from any plant material. One should be well aware of this fact, particularly when new plant material, which may contain unknown gibberellins, is to be analyzed. Secondly, the procedures employed should be adapted to the purpose of the study, i.e., identification and structure determination require a high degree of purity and in this case relatively large amounts of plant material are extracted and rigorously purified at the expense of great losses before analytical methods can be employed.

In contrast, for quantitative determinations of gibberellin activity in extracts, usually studied in relation to certain developmental processes, in general relatively small samples of material are available. In the latter case losses have to be limited, and one must be content with less rigorous identification, but instead place more emphasis on sensitivity, reproducibility, and speed of analysis.

3 Group Separation Procedures

Before highly separative analytical techniques can be employed, the acidic ethyl acetate soluble, as well as other fractions obtained after partioning, require further purification by chromatographic procedures with high sample capacity. This is necessary because of the low hormone content of plant extracts relative to the high dry weights of the sample. Another reason for using group separatory procedures is the wide diversity of gibberellins and their conjugates. In addition, it is often necessary to separate the gibberellins as a group from the numerous other low molecular compounds present in extracts. In the course of time several chromatographic procedures have been used.

3.1 PVP Adsorption Chromatography

With this method, as much as 200 mg extract residues are loaded onto a column (300 × 19 mm ID) filled with wet polyvinylpolypyrrolidone (PVP). The gibberellins are eluted between 60 and 160 ml with 0.1 M phosphate buffer, pH 8.0, as mobile phase. In this way sample weights are reduced substantially, and the recovery of gibberellins is reported to be more than 90% (Glenn et al. 1972). The technique is flexible and uses an inexpensive support and, of course, varying column diameters can be used to accommodate a wide range of sample sizes.

Alternatively, a PVP slurry technique can be used by adding PVP to the aqueous phase after partitioning with petroleum ether (Crozier and Durley 1983).

3.2 Gel Permeation or Steric Exclusion Chromatography

This method was used by Reeve and Crozier (1976, 1978) as a preparative group separatory procedure. Their system consists of two columns (1000 × 25 mm ID) connected in series, packed with Biobeads SX4 and eluated with tetrahydrofuran at a flow rate of 2 ml min^{-1}, this being the maximum flow rate the support can tolerate without excessive compression of the bed. The gel has an operating range of molecular weight 0–1500 and solutes elute in decreasing order of molecular size. This system has a high sample capacity with recoveries greater than 90%. The absence of absorption effects ensures that even with very impure extracts all solutes will be eluted within a solvent volume not exceeding the volume of the column. This allows the system to be used repeatedly without overlap. GA_9 and GA_{43}, which represent the extremes of the molecular weight range of the free GA's, elute within a 100–120 ml volume in this system (Reeve and Crozier 1976).

3.3 Anion Exchange Chromatography

Gräbner et al. (1976) have used DEAE Sephadex A-25 anion exchange chromatography to separate abscisic acid, GA_3, GA_7, and the 3-O-β-D glucosyl ether of GA_3. This procedure is readily adapted for use as a group separatory procedure for free gibberellins. DEAE Sephadex A-25 is loaded onto a column (150 × 40 mm ID), charged in the acetate form, and eluted with four void volumes (600 ml) of methanol-0.2 N acetic acid (1 : 1, v/v) to remove neutral and weakly acidic impurities.

The gibberellins are subsequently eluted at about 90% efficiency with two void volumes of methanol-2.0 N acetic acid (1 : 1, v/v).

3.4 Charcoal Adsorption Chromatography

Charcoal adsorption chromatography shows unusual reversed phase effects which can be employed to purify gibberellins in extracts (Yabuta and Hayashi 1939). Samples dissolved in 1–2 ml of 20% aqueous acetone are applied to the top of a charcoal celite (1 : 2, w/w) column (120 × 20 mm ID) and the weakly adsorbed impurities eluted with about 100 ml of 20% aqueous acetone (equal to 4 column volumes). The gibberellins are then removed with acetone (200 ml). The sample capacity of charcoal is high and the column size is easily adapted for extract residues weighing up to 500 mg. Charcoal can also be used as a slurry procedure like PVP, in which case large numbers of extracts can be treated rapidly. The recovery of gibberellins from charcoals is usually 75–85%; however, occasionally high losses may occur without explanation. This unreliability can be overcome by using replicates and internal standards.

3.5 Sephadex G-10 Chromatography

This method has been used by Crozier et al. (1969) for purification of gibberellins. A Sephadex G-10 column is eluted with 0.1 M phosphate buffer, pH 8.0, which retains gibberellins by uncharacterized adsorption phenomena. The procedure, however, is time-consuming and less effective than other procedures.

3.6 Countercurrent Distribution

This method was used in the early days for gibberellin purification by Crozier et al. (1969) among others, but has found no wide application. Countercurrent distribution is relatively very time-consuming.

The above-described group separatory procedures are mainly used for relatively large-scale purification for the purpose of identification of gibberellins. These methods are time-consuming, involve relatively large volumes of solvents, and are thus hardly suitable for routine gibberellins analysis, which requires rapid analysis and handling of many samples at a time. Recent developments with regard to disposable extraction columns for prepurification of plant extracts which combine small volumes of eluting solvents with high speeds of analysis meet these criteria.

3.7 Silica Gel Coated Disposable Extraction Columns

Small disposable columns (volume 1–6 ml) containing silica gel or silica gel coated with chemically bonded octadecyl (C_{18}) silane or octyl (C_8) cyano (CN), phenyl (C_6H_5), etc. are now becoming widely used for prepurification of extracts to be analyzed with HPLC.

For gibberellins C_{18} Sep-Pak cartridges (Millipore Waters) have been used successfully for purification of relatively crude plant extracts, e.g., after partitioning with petroleum ether (Fig. 2) (Barendse et al. 1980; Crozier and Durley 1983).

As a first step, about 10 ml of 70% aqueous methanol in 20 mM ammonium acetate buffer, pH 3.5, is forced through the Sep-Pak cartridge with a syringe. The extract dissolved in 5 ml 70% methanol is subsequently also forced through the cartridge, followed by further elution with another 5 ml of 70% methanol. Under these conditions many of the less polar contaminants are retained, while the gibberellins pass through the cartridge and are recovered from the methanolic eluate in more than 90% yields.

The sample capacity of Sep-Pak cartridges is low, and overloading results in bad separation and purification as well as low recovery. Baker Chemicals have introduced also a wide range of disposable extraction columns which have a choice of different packings, which come in different sizes. In principle these columns allow even better separation and purification, when different columns are combined, as well as higher sample capacities. The experience with these columns is, however, still limited with regard to gibberellin purification.

4 High Performance Liquid Chromatography (HPLC)

Gibberellins in plant extracts have been analyzed by both preparative and analytical, often both in combination, HPLC. The preparative HPLC increasingly replaces the elaborate and time-consuming group separatory procedures described above. This has become possible with the advent of a new generation of columns which combine relatively high sample capacities and powerful separation capabilities.

4.1 Gel Permeation or Steric Exclusion HPLC

Steric exclusion chromatography has not found widespread application, primarily because of the large size and instability of gel columns. However, with the advent of macropore cross-linked styrene-divinyl-benzene copolymer supports, specifically designed for HPLC, the procedure has been simplified and may be as rapid and convenient as any other HPLC procedure (Crozier et al. 1980). Gel permeation or steric exclusion separates on molecular size rather than polarity. This difference in mechanism will enhance the overall effectiveness of gibberellin purification in extracts, particularly when combined with analytical HPLC separating on different characteristics of gibberellins, e.g., polarity.

Steric exclusion HPLC involves the use of a 10 µm PL gel support (Polymer Laboratories Ltd., formerly μ-spherogel from Altex Scientific Inc. USA) with a nominal exclusion limit of less than 2000 molecular weight. The column (300 × 10 mm ID) is eluted with 0.5% acetic acid in tetra hydrofuran at a flow of 1 ml min^{-1}. In this system 9000 theoretical plates are generated, with a sample capacity of more than 100 mg. Crozier et al. (1980) have used this technique to separate the methoxy coumaryl esters of free gibberellins, prepared according to the procedures described by Durst et al. (1975) and Dünges (1977), and detected

Table 1. Steric exclusion chromatography (SEC) characteristics of gibberellin methoxycoumaryl esters. (Crozier et al. 1980)

Compound	Mol. wt.	Solvent[a]	Retention volume (ml)
$GA_{13}CE$	942	1, 2	6.3
$GA_{25}CE$	926	1, 2	6.3
$GA_{36}CE$	738	1, 2	6.3
$GA_{14}CE$	724	1, 2	6.3
$GA_{8}CE$	552	1, 2	6.3
$GA_{16}CE$	536	1, 2	6.5
$GA_{1}CE$	536	1, 2	6.4
$GA_{3}CE$	534	1, 2	6.4
$GA_{4}CE$	520	1, 2	6.5
$GA_{20}CE$	520	1, 2	6.5
$GA_{5}CE$	518	1, 2	6.5
$GA_{7}CE$	518	1, 2	6.5
$GA_{9}CE$	504	1, 2	6.8
GA_{43}	394	2	7.0
GA_{9}	316	2	7.6

[a] Solvent 1 = tetrahydrofuran (THF); solvent 2 = 100 mM acetic in THF.

by spectrophotometry. This technique provides only partial resolution of individual gibberellin coumaryl esters, and in fact represents a group separatory procedure of preparative HPLC.

The elution characteristics of a number of gibberellin coumaryl esters separated by steric exclusion chromatography are compiled in Table 1.

The preparation of the methoxy coumaryl esters of gibberellins allows spectrophotofluoremetric detection of these hormones at the pg level and the method will be described in more detail in Sect. 4.4. Crozier et al. (1980) used this technique also with radioactive gibberellins employing a radioactivity monitor.

In principle, one can use this technique also as a group separatory procedure of nonderivatized free gibberellins using radioactive internal standards for reference.

4.2 Normal Phase HPLC

Crozier and Reeve (1977) described a preparative normal phase HPLC procedure for the separation of different free gibberellins and precursors. They used columns (450 × 10 mm ID) containing either Partisil 10 or Partisil 20 coated with a 0.5 M formic acid stationary phase, which were eluted with a gradient of increasing amounts of ethyl acetate in hexane. The samples, dissolved in mobile phase, were introduced off-column via a six-port valve with a 130 μl loop. The column effluent was first passed through a UV monitor and then to an onstream radioactivity monitor (Reeve et al. 1976). In order to obtain sensitive UV detection, gibberellin benzyl esters were prepared from dried samples treated with a solution of dimethyl formamide dibenzylacetal in dry dioxane for 2 h at 70 °C. In this way the conversion of gibberellins to their respective benzyl esters was 80–90% effective for both standards and extracts which are subsequently analyzed by HPLC. The gibberellin benzyl esters of plant extracts are separated from the excess derivatizing reagents by preparative HPLC on a Partisil 10 column (450 × 10 mm ID) with a stationary phase of 40% 0.5 M formic acid and a mobile phase consisting of a gradient of 80–100% ethyl acetate in hexane. The appropriate collected fractions are then further separated and purified by analytical HPLC on a (500 × 4.6 mm ID) Partisil 10 column with a mobile phase of tetrahydrofuran dichloromethane (4:96, v/v). The fractions containing the gibberellin benzyl esters can then be further analyzed by mass spectrometry, as has been described by Crozier and Reeve (1977) for the ethyl acetate-soluble extract from *Phaseolus coccineus* seedlings made 24 h after application of ^{3}H-GA_4. Crozier and Reeve (1977) noted that the combined effects of preparative and analytical HPLC resulted in a 4000- to 10,000-fold purification of the gibberellins studied with recoveries of 20–30%. Since in their case the levels of gibberellins in the ethyl acetate soluble fraction were relatively high, it may be necessary in the case of much lower concentrations of gibberellins in the extract to apply additional analytical HPLC before the samples are suitable for GC-MS analysis.

The described preparative normal phase HPLC can also be applied to nonderivatized gibberellins, which is shown by the use of radioactive gibberellins and gibberellin precursors in Fig. 3.

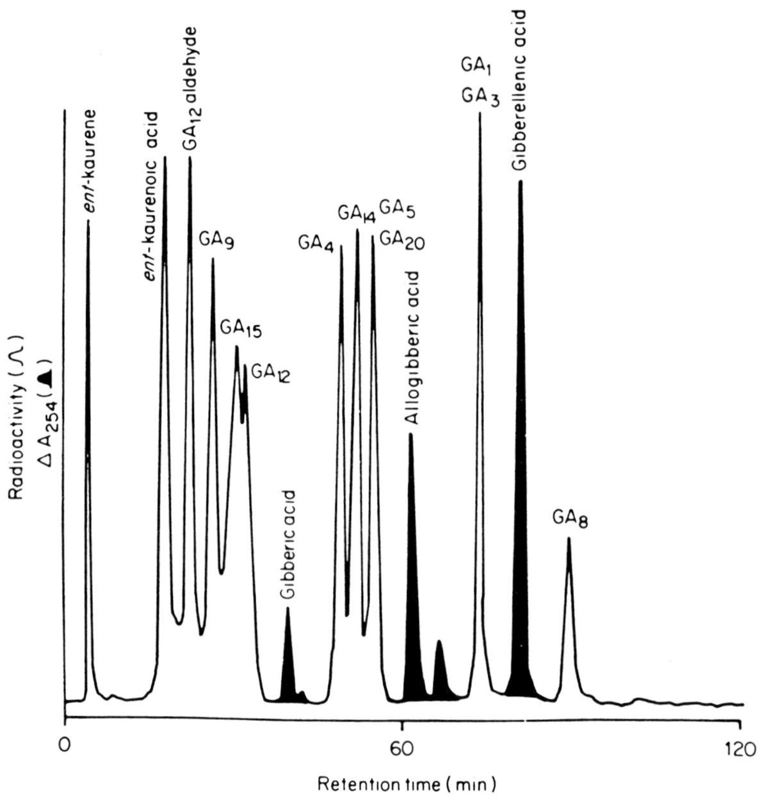

Fig. 3. Preparative HPLC of radioactive GA's and GA precursors with UV-absorbing internal markers. Column: 450 × 10 mm ID Partisil 20. Stationary phase: 40%, 0.5 M formic acid. Mobile phase: 2-h gradient 0–100% ethyl acetate in hexane. Flow rate: 5 ml min^{-1}. Sample: ca. 24,000 dpm ent-[^{14}C]kaurene, 50,000 dpm [^{14}C]GA_3, [^{3}H]GA_5, [^{14}C]GA_{12}, [^{14}C]GA_{19}, and [^{3}H]GA_{20}; 100,000 dpm ent-[^{3}H]kaurenoic and [^{3}H]GA_1, [^{3}H]GA_4, [^{3}H]GA_8, [^{3}H]GA_9, [^{3}H]GA_{12} aldehyde, and [^{3}H]GA_{14} and uncalibrated amounts of gibberic acid, allogibberic acid, and gibberellenic acid. Detectors: homogeneous radioactivity monitor, 1800 cpm full-scale deflection, absorbance monitor at A 254. (Reeve et al. 1976)

4.3 Reversed Phase HPLC

Fractionation of gibberellins by reversed phase HPLC is becoming the method of choice, and this technique is at present most widely used. Jones et al. (1980) developed a combined preparative and analytical reversed phase HPLC procedure for the fractionation of gibberellins in plant extracts. Their preparative HPLC system consisted of a series of four stainless steel columns (each 60 × 6.5 mm D) packed with Bondapak C_{18}/Porasil (Millipore Waters). For elution two solvents were used, i.e., 1% aqueous acetic acid (A) and 95% ethanol (B) run in a linear gradient of B in A, respectively 30 to 100% in 25 min at a flow rate of 9.9 ml min^{-1}. The samples were dissolved in 70% A plus 30% B, filtered through 0.45-µm type HA Millipore filters and the gradient was started 30 s after injection. In this way the authentic gibberellins used were eluted in groups over at least two, but usually three 1-min fractions, starting with the most polar gibberellins. The main factor determining elution time was degree and position of the hydroxyl groups on the gibberellin molecules.

This procedure was applied to an acidic extract, prepurified by charcoal/celite column chromatography, of *Pharbitis nil* seeds (Jones et al. 1980). A part of each fraction collected from this preparative HPLC was tested for biological activity with the dwarf corn d_5 assay. The fractions showing biological activity were subjected to analytical HPLC on a μ-Bondapak C_{18} column using a linear gradient of methanol (30–100% or 10–70%) in 1% aqueous acetic acid, run in 30 min at 2 ml min^{-1}. One-minute fractions were collected and 1/50 of each fraction was again assayed with the d_5 corn bioassay. The various biological fractions were then methylated and silylated for identification by GC-MS.

In this way Jones et al. (1980) not only confirmed the presence of five known gibberellins in *Pharbitis* seeds, but also two new gibberellins were identified.

The quantity of material used in this analysis was 16 g dry weight, which is considerably less than the amounts used by earlier workers, which were in at least in the kg range. However, seeds generally contain fewer impurities than extract from vegetative tissues which often requires additional steps of purification.

This reversed phase HPLC procedure also allows the isolation and separation of other acidic plant growth substances, e.g., abscisic acid, indole acetic acid, dihydrophasic acid, and probably also of gibberellin conjugates.

The above reversed phase HPLC procedure is also combined with gas-chromatography-selected ion current monitoring for quantitative analysis of gibberellins in plant extracts, as has been described by Metzger and Zeevaart (1980) for gibberellins in spinach, as well as for the identification and quantitation of metabolites following treatments of plants with [^{3}H] GA_{20} (Metzger and Zeevaart 1982). Koshioka et al. (1983) described the fractionation of both acidic and conjugated gibberellins by reversed phase HPLC using Water Assoc. μ-BondapakC$_{18}$ column (300 × 3.9 mm ID) and Radial-PAK A cartridge C_{18} (100 × 8 mm ID), and a Whatman Partisil M9 10/50 ODS-2 (500 × 9 mm ID) column. The elution solvents consisted of aqueous 1% acetic acid and methanol, which were used in a combination of linear gradients with time. In this manner a number of free gibberellins, some of their methyl esters, precursors such as kaurene, kaurenoic acid, and GA_{12}-aldehyde, including some related compounds such as gibberellenic acid, epi-GA_1, GA_{29} catabolite, and others, as well as other plant hormones, e.g., abscisic acid, were more or less separated. Detection of these various compounds was by UV monitor at 254 nm, fluorescence monitor (λ^{ex} 285 nm, λ^{em} 370 nm), radioactivity monitor and bioassay. Barendse et al. (1980) used reversed phase HPLC on Radial PAK-A C_{18} cartridges to separate nonderivatized gibberellins, as will be described in more detail in the following section.

Yamaguchi et al. (1979) have used two kinds of reversed phase column, octadesylsilanized (Wakogel ODS) and dimethylsilanized silica gel (Merksorb S160) for the analysis of conjugated gibberellins from immature seeds of *Quamoclit pennata*.

They used different percentages of methanol in ammonium chloride (pH 3.2) under isocratic conditions, and the conjugated gibberellins were monitored at 200–210 nm.

4.4 Detection of Gibberellins After HPLC

One of the main problems in using HPLC for gibberellin analysis is their detection. Gibberellins have very low UV absorption and can be detected only in the lower wavelength range, i.e., 203–210 nm, at reasonable sensitivity (μg level) (Barendse et al. 1980). In order to enhance UV detection, Reeve and Crozier (1978) chose to derivatize the gibberellins by esterification of the carboxyl group at C-7. The resulting gibberellin benzyl esters have an absorption maximum at 256 nm and can thus be detected with a standard UV monitor at 254 nm. Morris and Zaerr (1978) prepared the 4-bromophenacyl esters of gibberellins for separation of gibberellins by HPLC.

The HPLC separation of double bond isomers of gibberellins, such as GA_1 and GA_3, GA_4 and GA_7, GA_5 and GA_{20}, was achieved by Heftmann et al. (1978) in the form of their p-nitrobenzyl esters.

However, these HPLC separations with derivatized gibberellins were carried out with authentic gibberellins and limited experience is available with regard to

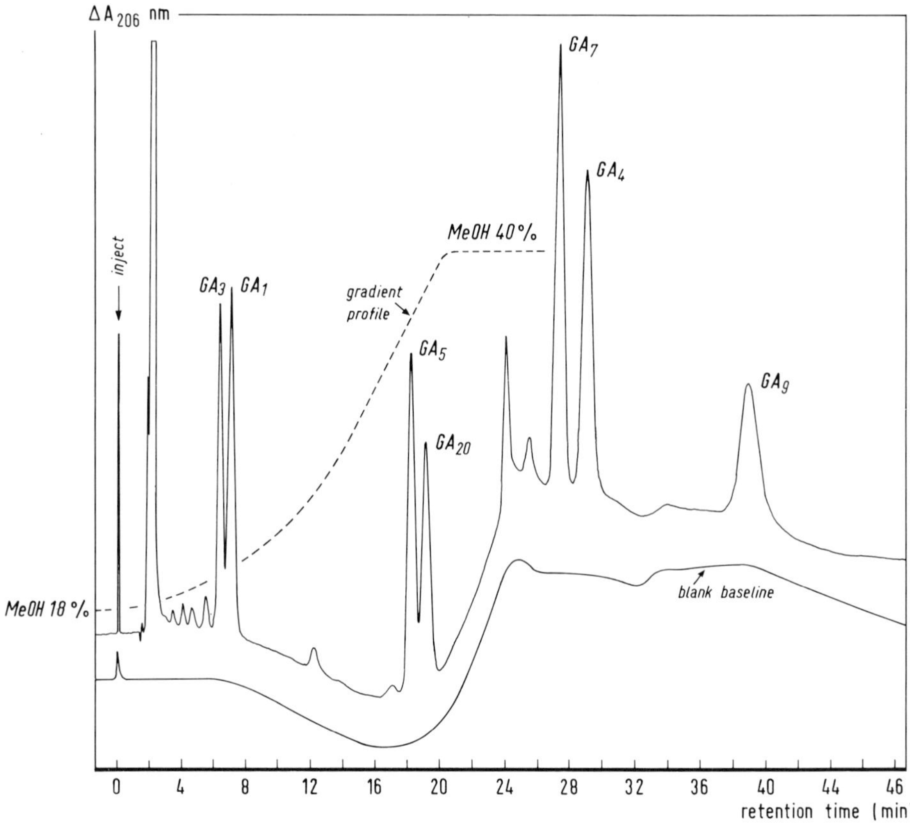

Fig. 4. Reversed phase HPLC of GA's. Column: Bondapak C_{18}, 300 × 39 mm ID. Mobile phase: gradient from 18% methanol, containing 0.01 M H_3PO_4 and acetone (10 ml l^{-1}) to 40% methanol 0.01 M H_3PO_4 at pH 2.2 in 20 min. Flow rate: 1.8 ml min^{-1}. Detection at 206 nm. (Barendse et al. 1980)

such analyses of plant extracts. We have found (Barendse et al. 1980) the derivatization of gibberellins in plant extracts rather troublesome and unreliable, particularly when gibberellins are present in minute quantities as in vegetative material, which contains many impurities which are also derivatized and thus obscure the analysis. For the same reasons, Zeevaart and coworkers did not derivatize their plant extracts (Jones et al. 1980; Metzger and Zeevaart 1980), but instead employed bioassays to detect the active fractions collected after HPLC. However, in this way only biologically active gibberellins are detected.

Barendse et al. (1980) developed a reversed phase HPLC procedure which enables good separation, including of the gibberellin double bond isomers GA_1 and GA_3, GA_4 and GA_7, GA_5, and GA_{20}, without derivatization. The reversed phase HPLC carried out on a μ-Bondapak C_{18} column with 20% methanol in 0.01 M H_3PO_4, adjusted to pH 2.3 as mobile phase, achieved the separation of the double bond isomers GA_1 and GA_3. Three pairs of double bond isomers could be sparated in one run with a gradient from 18% methanol in 0.01 M H_3PO_4 and acetone (10 ml l^{-1}) to 40% methanol in 0.01 M H_3PO_4 at pH 2.2 in 20 min. Detection took place at 206 nm (Fig. 4).

These separations were further improved by using a Radial Pak-A column (Millipore Waters), which has the additional advantage that one does not have to resort to a methanol gradient which causes baseline shifts but merely to apply a pH gradient (Fig. 5).

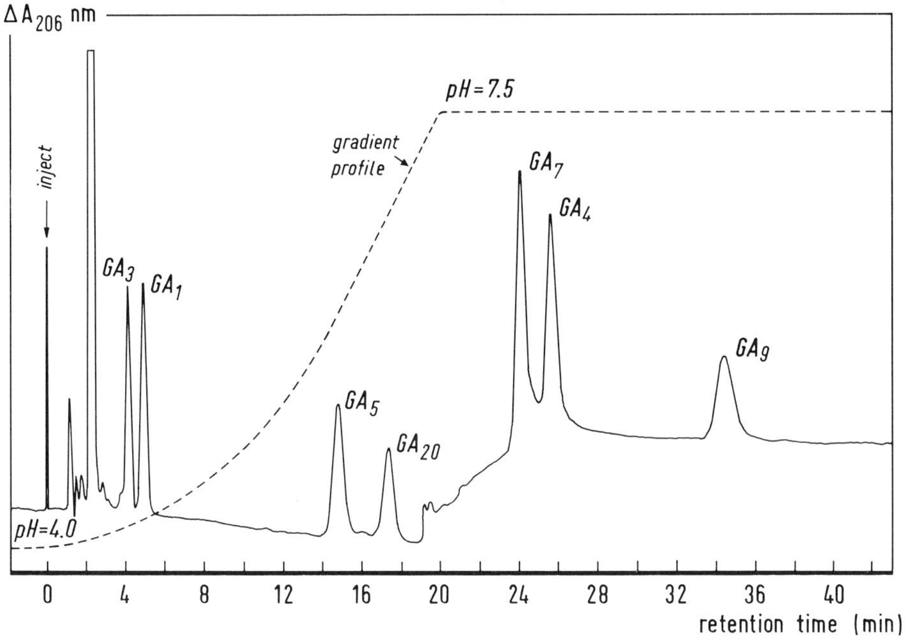

Fig. 5. Reversed phase HPLC of GA's. Column: Radial-PAK-A. Mobile phase: gradient from 35% methanol, 0.01 M H_3PO_4 at pH 4.0 to 35% methanol, 0.01 M H_3PO_4 at pH 7.5 in 20 min. Flow rate: 2 ml min^{-1}. Detection at 206 nm. (Barendse et al. 1980)

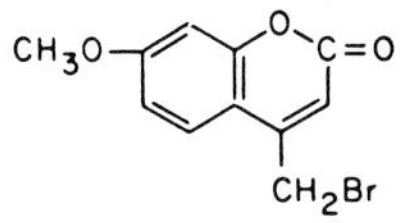

4-bromomethyl-7-methoxycoumarin (BMC)

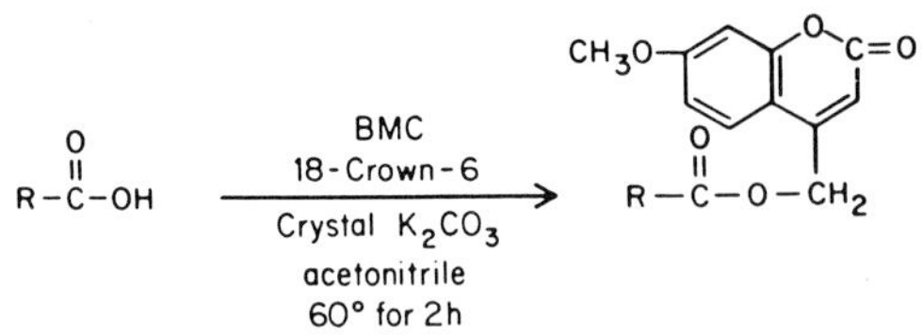

Fig. 6. Crown ether catalyzed synthesis of methoxycoumaryl esters. (Crozier et al. 1982)

By choosing a particular methanol concentration (in 0.01 M H_3PO_4) and an appropriate pH, a particular desired group of gibberellins can be analyzed with good resolution. In addition, the solvent system can be adapted to separate conjugated gibberellins, as was done for the GA_3 and GA_{20} glucosyl ethers (Barendse et al., unpublished data). In this system the gibberellins are separated mainly according to their degree of hydroxylation, the more hydroxyl groups the shorter the retention time. The above reversed phase HPLC procedure was also used for analyzing plant extracts using bioassay or internal radioactivity gibberellin standards for detection of gibberellin-containing fractions (Barendse et al. 1980, 1983). Lin and Heftmann (1981) improved the method of Barendse et al. (1980) by substituting the H_3PO_4 in the mobile phase with 5% formic acid. At present we prefer to use 0.01 M acetic acid instead of H_3PO_4 in the mobile phase.

More recently Crozier et al. (1982) reported the reversed and normal phase HPLC of gibberellin methoxycoumaryl esters which are strongly fluorescent derivatives (λ_{max}^{exit} 320 nm, λ_{max}^{emis} 400 nm) which can be monitored with a fluorescence spectrometer. This enables detection of gibberellins at the picogram level in contrast to the above described methods, where at best 100 ng or higher can be detected. The methoxycoumaryl esters are prepared according to Dünges (1977) by reacting the free gibberellin acids with an equimolar amount of 4-bromomethyl-7-methoxyl coumarin, an one-tenth molar equivalent of 18-crown-6 and a crystal of K_2CO_3 in 100 µl of dry acetonitrile at 60 °C for 2 h (Fig. 6).

The reaction mixture is then evaporated to dryness, water added, the gibberellin coumaryl esters extracted into chloroform and the samples dried prior to purification by steric exclusion and reversed phase HPLC. Reversed phase and normal phase HPLC separations of the gibberellin coumaryl esters were carried out on respectively a 5 µm ODS-Hypersil column (250 × 5 mm ID) eluted with varying ratios of either methanol or ethanol in 20 mM ammonium acetate buffer (pH 3.5), and a 5 µm CPS-Hypersil column eluted isocratically with various ratios of hexane and dichloromethane containing 3% ethanol. The column effluent was monitored with a spectrophotofluorimeter (excitation 320 nm, emission 400 nm) and the radioactive compounds were also concurrently detected with a radioactive monitor. The fluorescence detection limit obtained by Crozier et al. (1982) is illustrated in Fig. 7. The limit of GA_3 detection, which forms a mono-ester, is ca. 1 pg (2.8 fmol), as determined by the point at which the curve intersects the

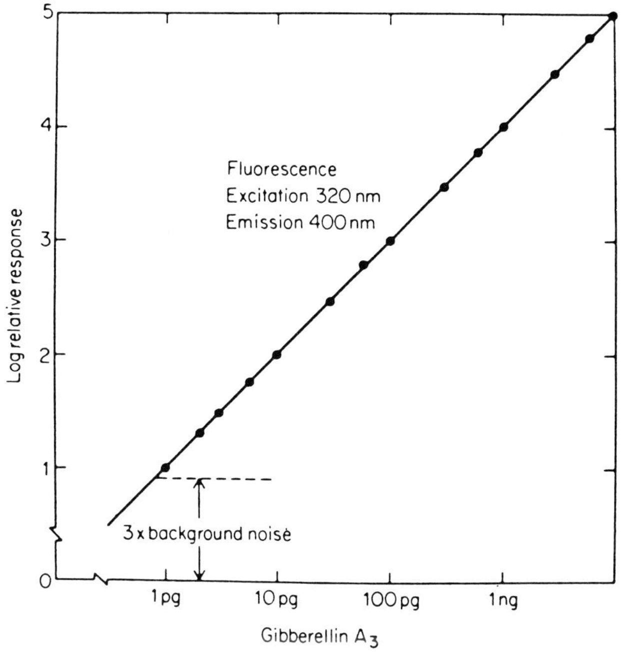

Fig. 7. Fluorescence detection limit of GA_3CE after reverse-phase HPLC. Column: 250 × 5 mm ID ODS Hypersil. Mobile phase: 45% ethanol in 20 mM, pH 3.5 ammonium acetate buffer. Flow rate: 1 ml min^{-1}. Sample: GA_3CE ($k'=2.3$) load as indicated. Detector: Perkin-Elmer 650-10 LC spectrophotofluorimeter, excitation 320 nm, emission 400 nm. (Crozier et al. 1982)

ordinate equivalent to three times the level of background noise. With bis- and tris derivatives the value will be correspondingly lower.

Reversed phase HPLC separation of a range of gibberellin coumaryl esters obtained by gradient elution from 40–80% ethanol in 20 mM ammonium acetate buffer (pH 3.5) from a ODS-Hypersil column (250 × 5 mm ID) is illustrated in Fig. 8.

This system is able to separate the coumaryl esters of closely related gibberellins such as the double bond isomers GA_1 and GA_3, GA_4 and GA_7, GA_5 and GA_{20}, as well as other closely related gibberellins, i.e., GA_{13} and GA_{14}, GA_9 and GA_{36} (Crozier et al. 1982).

In general, an increasing number of free hydroxyl groups decreases retention time, 13α-hydroxylation to a greater extent than 3β-hydroxylation, which in turn is more effective than hydroxylation at either the 1α or 2β position.

In addition the elution order (Fig. 8) shows a retention order of mono < bis < tris esters, indicating that increasing the number of methoxycoumaryl functions decreases polarity.

In a similar manner (Crozier et al. 1982), normal phase HPLC separation was carried out on a CPS-Hypersil column eluted isocrationally with 3% ethanol in either 12% or 20% dichloromethane in hexane. Although the effects of 3β and 13α-hydroxylation and 1,2 and 2,3 double bonds are, as anticipated, the opposite

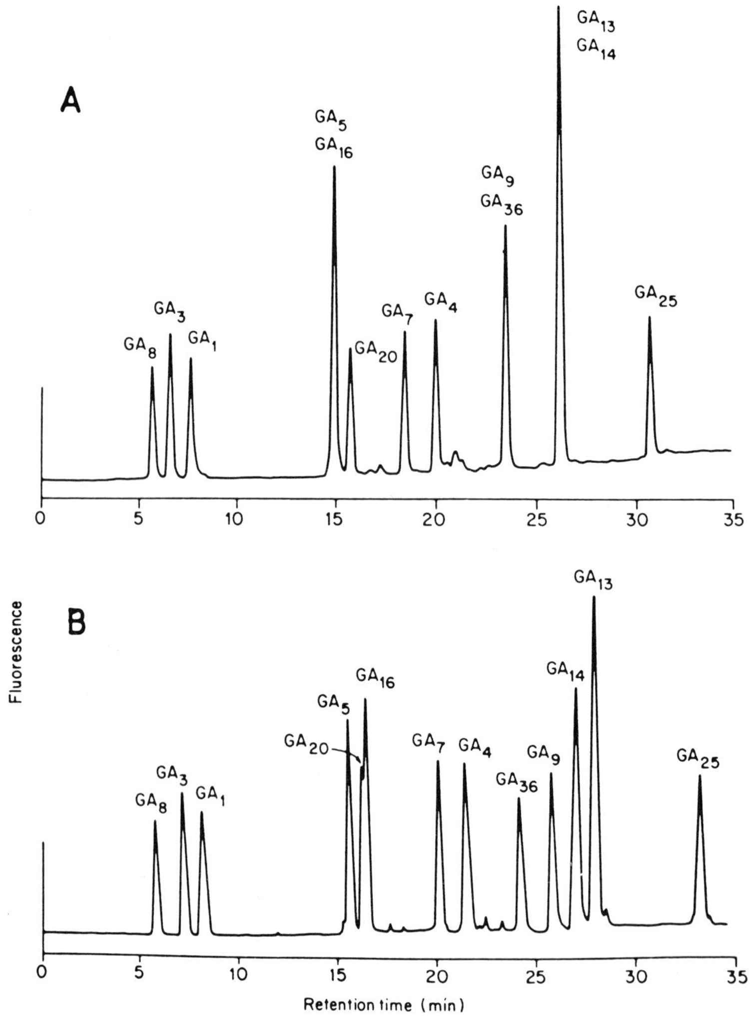

Fig. 8. Reverse-phase HPLC of GACE's. Column: 250 × 5 mm ID ODS Hypersil. Mobile phase: 30-min gradient: **A** 60–100% methanol in 20 mM, pH 3.5 ammonium acetate buffer; **B** 40–80% ethanol in 20 mM, pH 3.5 ammonium acetate buffer. Flow rate: 1 ml min^{-1}. Sample: methoxycoumaryl esters of GA_1, GA_3, GA_4, GA_7, GA_8, GA_9, GA_{13}, GA_{14}, GA_{16}, GA_{20}, GA_{25}, and GA_{36}. Detector: Perkin-Elmer 650-10LC spectrophotofluorimeter excitation 320 nm, emission 400 nm. (Crozier et al. 1982)

of those observed with reversed phase HPLC, the overall elution pattern is not a mirror image of the ODS-Hypersil profile of Fig. 8.

Gibberellin coumaryl esters are not readily volatile and as a consequence mass spectra have to be obtained by direct-probe mass spectrometry rather than by GC-MS.

The high sensitivity of fluorescence detection of gibberellin coumaryl esters enhances the overall flexibility of HPLC as an analytical tool for trace amounts of endogenous gibberellins. However, this high sensitivity is not accompanied by an increase of detector selectivity. On the contrary, gibberellins do not exhibit native fluorescence, and when plant extracts are derivatized, not only the gibberellins but also any carboxylic acid in the extract will form methoxycoumaryl esters. As a consequence this method requires extensive prepurification before an accurate and reproducible analysis is achieved, as was shown by Davies et al. (cited after Crozier and Durley 1983) for the detection of GA_1 in an extract of *Salix pentandra* shoots. Barendse (unpublished) had similar experiences with extracts from *Pharbitis nil* cv. violet seeds.

In addition, it was established that the lowest amount of authentic GA_3 which could be derivitized into its coumaryl ester was approximately 1 ng per sample. The high sensitivity of gibberellin detection at the picogram level reported by Crozier et al. (1982) was obtained by dilution of the sample after derivatization.

We found, on adding known amounts of authentic GA_3 to plant extracts, that even much more than 1 ng has to be present in the sample in order to be detected by fluorescence.

This means that, together with the high degree of purity required, the sensitivity of the method is not as great as it appears at first sight. Nevertheless, we found this method of Crozier et al. (1982) very useful when the prepurification was carried out with reversed phase HPLC of nonderivatized gibberellins as described by Barendse et al. (1980). In this case the selectivity of the system is also enhanced significantly.

5 Recent Developments and Prospects

The great advantage of HPLC in the analysis of plant hormones in general, and of gibberellins in particular, lies in its powerful separation capabilities. HPLC is comparatively rapid and normally does not require the rigorous prepurification procedures necessary for GC-MS analysis. There is an ever-extending choice of column packings available which allow the most complicated and difficult purification and separation of plant extracts to be solved, e.g., by combining different column types and various solvent systems for elution. In this respect particular reversed phase HPLC appears to become the method of choice. In addition HPLC is nondestructive and therefore may be readily combined with other analytical procedures such as GC-MS, or may in future be directly coupled to LC-MS.

HPLC systems are now available in many degrees of sophistication which makes HPLC affordable for most laboratories.

A major disadvantage of HPLC at the moment is its low selectivity due to the general use of UV monitors for detection. Although other types of detector are available for HPLC, e.g., refractive index, fluorescence, and electrochemical detectors, and more recently the radioactivity monitor, these were either not suitable or not sensitive enough for plant hormone detection. However, recent developments with regard to enhanced sensitivity of detection or increased selectivity of HPLC have enhanced the analytical capabilities of HPLC.

5.1 Diode Array UV Detection

HPLC has so far lacked the availability of a universal UV detector which allowed optimizing the response, detection limits, and resolution. Continuous data can at present usually only be obtained at one preselected wavelength, which limits the amount of information to be obtained.

In addition, the traditional spectrophotometric detectors can only scan under stopped-flow status, which is tedious and time-consuming, while it also leads to band broadening, adversely affecting separation, and virtually precludes quantitation.

These difficulties have now been overcome by the recent advent of the photodiode array-detector (LKB and Hewlett-Packard). This detector is based on a dynamic system using reversed optics. Light is projected through the flow cell, dispersed via a holographic grating onto a number of photosensitive diodes, resulting in voltage which is then converted to digital signals for further processing by a personal computer. Diode array detection enables continuous spectral determination during elution and allows rapid simultaneous access to data of the total preprogrammed wavelength range which are stored on a diskette. With this detector, widely different compounds can be detected simultaneously, and the spectra of each compound provide a maximum degree of selectivity. Recent experiences (Barendse, unpublished) with regard to the analysis of auxins, cytokinins as well as gibberellins, in plant extracts showed great promise for the application of combined HPLC diode-array detection for plant hormone analysis.

5.2 Electrochemical Detection

Although electrochemical detection is not new, a highly sensitive detector of this kind for HPLC has only recently become available. The electrochemical detector monitors changes in currents associated with the reduction and/or oxidation of sample components, which allows for highly sensitive and selective detection. This type of detection has become extremely sensitive and is capable of rapid analysis of a wide range of compounds at or below the pg level. It has been used successfully for the detection of auxin, i.e., indolyl acetic acid (personal communication).

5.3 Combined HPLC Immunoassay

Immunoassays represent at the moment the most sensitive techniques to detect plant hormones, including gibberellins, at the fmol level in plant extracts. Besides being very hormone-specific, immunoassay requires only small volumes of semipurified extracts. Thus the combination of the powerful HPLC separation capabilities with the very high sensitivity and selectivity of immunoassays provides an attractive alternative with regard to routine quantitative plant hormone analysis.

MacDonald et al. (1981) described a combined HPLC-radioimmunoassay for cytokinins, and this technique was elaborated by Akiyoshi et al. (1983), who combined immuno-affinity chromatography and HPLC-radioimmunoassay to isolate and exactly measure the levels of cytokinins in plant tissue. There is no doubt that these techniques can also be used for gibberellins.

References

Akiyoshi DE, Morris RO, Hink R, Mischle BS, Kosuge T, Garfunkel DJ, Gordon MP, Nester EW (1983) Cytokinin/auxin balance in grown gall tumors is regulated by specific loci in the T-DNA. Proc Natl Acad Sci USA 80:407–411

Barendse GWM, Van der Werken PH, Takahashi N (1980) High performance liquid chromatography of gibberellins. J Chromatogr 198:449–455

Barendse GWM, Dijkstra A, Moore TC (1983) The biosynthesis of the gibberellin precursor *ent*-kaurene in cell-free extracts and the endogenous gibberellins of Japanese Morning Glory in relation to seed development. J Plant Growth Regul 2:165–175

Binks R, MacMillan J, Pryce RC (1969) Plant hormones VII. Combined gas chromatography-mass spectrometry of the methyl esters of gibberellins A_1 to A_{24} and their trimethylsilyl esters. Phytochemistry (Oxf) 8:271–284

Brenner ML (1981) Modern methods for plant growth substance analysis. Annu Rev Plant Physiol 32:511–538

Crozier A, Aoki H, Pharis RP (1969) Efficiency of counter-current distribution. Sephadex G-10 and silicic acid partition chromatography in the purification and separation of gibberellin-like substances from plant tissue. J Exp Bot 20:786–795

Crozier A, Durley RC (1983) Modern methods of analysis of gibberellins. In: Crozier A (ed) The biochemistry and physiology of gibberellins, vol 1. Praeger, Eastbourne, East Sussex, pp 485–560

Crozier A, Reeve DR (1977) The application of high performance liquid chromatography to the analysis of plant hormones. In: Pilet PE (ed) Proc 9th. Int Conf Plant Growth Subst. Springer, Berlin Heidelberg New York, pp 67–76

Crozier A, Zaerr JB, Morris RO (1980) High-performance steric exclusion chromatography of plant hormones. J Chromatogr 198:57–63

Crozier A, Zaerr JB, Morris RO (1982) Reversed- and normal-phase high-performance liquid chromatography of gibberellin methoxycoumaryl esters. J Chromatogr 238:157–166

Davies JR, Jensen E, Rivier L, Juntilla O, Crozier A: Detection of gibberellins in extracts from *Salix pentandra* shoots by high performance liquid chromatography and combined gas chromatography-mass spectrometry (cited after Crozier and Durley 1983)

Dünges W (1977) 4-Bromomethyl-7-methoxycoumarin as a new fluorescent label for fatty acids. Anal Chem 49:442–445

Durst D, Milano M, Katka GJ, Connely SA, Grushka E (1975) Phenacyl esters of fatty acids via crown ether catalysts for enhanced ultraviolet detection in liquid chromatography. Anal Chem 47:1797–1801

Glenn JL, Kuo CC, Durley RC, Pharis RP (1972) Use of insoluble polyvinyl pyrrolidone for purification of plant extracts and chromatography of plant hormones. Phytochemistry (Oxf) 11:345–351

Gräbner R, Schneider G, Sembdner G (1976) Gibberelline. 43ste Mitt. Fraktionierung von gibberellinen, gibberellinkonjugaten und anderen phytohormonen durch DEAE-Sephadex-chromatography. J Chromatogr 121:110–115

Heftmann E, Saunders GA, Haddon WF (1978) Argentation high-pressure liquid chromatography and mass spectrometry of gibberellin esters. J Chromatogr 156:71–77

Hiraga K, Yokota T, Murofushi N, Takahashi N (1974) Isolation and chracterization of gibberellins in mature seeds of *Phaseolus vulgaris*. Agric Biol Chem 38:2511–2520

Jones MG, Metzger JD, Zeevaart JAD (1980) Fractionation of gibberellins in plant extract by reverse phase high performance liquid chromatography. Plant Physiol (Bethesda) 65:218–221

Knox JH (1979) High performance liquid chromatography. Edinburgh Univ Press, Edinburgh

Koshioka M, Harada J, Takerro K, Noma M, Sassa T, Ogiyama K, Taylor JS, Road SB, Zegge RL, Pharis RP (1983) Reversed phase C18 high pressure/performance liquid chromatography of acidic and conjugated gibberellins. J Chromatogr 256:101–115

Lin J-T, Heftmann E (1981) Adsorption and reversed-phase partition high-performance liquid chromatography of gibberellins. J Chromatogr 213:507–510

MacDonald EMS, Akiyoshi DE, Morris RO (1981) Combined high-performance liquid chromatography-radioimmuno assay for cytokinins. J Chromatogr 214:101–109

Metzger JD, Zeevaart JAD (1980) Effect of photoperiod on the levels of endogenous gibberellins in spinach as measured by combined gas chromatography-selected ion current monitoring. Plant Physiol (Bethesda) 66:844–846

Metzger JD, Zeevaart JAD (1982) Photoperiodic control of gibberellin metabolism in spinach. Plant Physiol (Bethesda) 69:287–291

Morris RO, Zaerr JB (1978) 4-Bromophenocyl esters of gibberellins, useful derivatives for high performance liquid chromatography. Anal Lett A11:73–83

Reeve DR, Crozier A (1976) Purification of plant hormone extracts by gel permeation chromatography. Phytochemistry (Oxf) 15:791–793

Reeve DR, Crozier A (1978) The analysis of gibberellins by high performance liquid chromatography. In: Hillman JR (ed) Isolation of plant growth substances. Cambridge Univ Press, London, pp 41–78

Reeve DR, Crozier A (1980) Quantitative analysis of plant hormones. In: MacMillan J (ed) Hormonal regulation of development I. Molecular aspects of plant hormones. Springer, Berlin Heidelberg New York, pp 203–280

Reeve DR, Yokota T, Nask LJ, Crozier A (1976) The development of a high performance liquid chromatograph with a sensitive on-stream radioactivity monitor for the analysis of ^{3}H and ^{14}C labelled gibberellins. J Exp Bot 27:1243–1258

Takahashi N, Yamaguchi I, Yamane H (1986) Gibberellins. In: Takahashi (ed) Chemistry of plant hormones. CRC Press Inc., Boca Raton, Florida, pp 282, (ISBN-0-8493-5470-6)

Yabuta T, Hayashi T (1939) Biochemical studies on "bakanae" fungus of the rice II. Isolation of gibberellin, the active principles which makes the rice seedings grow slenderly. J Agric Chem Soc Jpn 15:257–266

Yamaguchi I, Yokota T, Yoshida S, Takahashi N (1979) High pressure liquid chromatography of conjugated gibberellins. Phytochemistry (Oxf) 18:1699–1702

Characterization of Cytokinins and Related Compounds by HPLC

CHONG-MAW CHEN

1 Introduction

1.1 Background

In recent years, high performance liquid chromatography (HPLC) has emerged as the chief method for the instrumental analysis of plant growth substances. It is a valuable tool for the separation, identification, and measurement of plant hormones and their metabolites. The combination of rapid analysis, high resolution and, frequently, the elimination of a derivatization step gives HPLC a marked advantage over other methods of separation including paper, thin-layer and gas chromatographies. For example, the use of HPLC for the analysis of cytokinins and their derivatives is superior to the classical technique of gas chromatography, since the liquid mobile phase allows the separation and recovery of cytokinins and their derivatives which are not readily volatilized. In addition, HPLC is to be preferred for molecules which have high polarity, a number of ionic groups, or thermal instability. Cytokinins and their derivatives purified by HPLC are often sufficiently pure for immunoassay or direct introduction into the mass spectrometer as trimethylsilated or permethylated derivatives, since the step using gas chromatography is unnecessary.

In some situations HPLC may substitute for the more time-consuming Sephadex LH-20 chromatography; however, there are many instances where the two techniques should be considered complimentary. Sephadex LH-20 columns often provide an initial purification step when large sample sizes make immediate preparative HPLC impractical.

Several excellent general texts have been published on HPLC (examples: Hancock and Sparrow 1984; Engelhardt 1985), which discuss the general concepts and techniques that are quite applicable to the analysis of plant hormones. Relevant articles concerning cytokinin analysis have recently been reviewed (Brenner 1981; Letham and Palni 1983; McGaw et al. 1984; Kashimizu and Iwamura 1986). This article will limit itself to a discussion of the variables involved in developing an HPLC analysis for cytokinins and related compounds.

1.2 Structure and Physicochemical Properties of Cytokinins and Related Compounds

The chromatographic behavior of a molecule or compound is dictated by its structure and its physicochemical property. The physicochemical behavior of cytokinins and their derivatives is a function of the purine ring structure, the side

Table 1. Structure of some cytokinins and related compounds

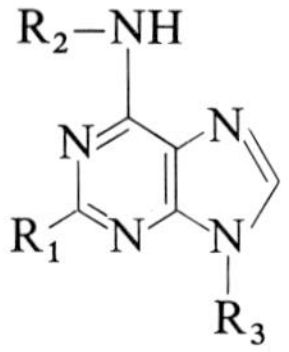

Substituents			Common name	Abbreviation
R_1	R_2	R_3		
H		H	N^6(Δ-isopentenyl)adenine or isopentenyladenosine	I^6Ade
H	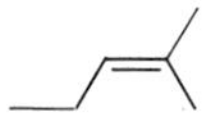	Ribosyl	N^6(Δ^2-isopentenyl adenosine) or isopentenyl-adenosine	i^6Ado
H	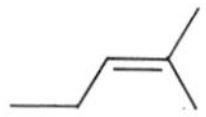	Glucosyl 9-glucoside	Isopentenyladenine	i^6Ado-9G
CH_3S		Ribosyl	2-methylthio-isopentenyl-adenosine	ms^2i^6Ado
H	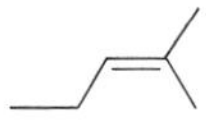	5′-phospho-ribosyl	Isopentenyladenosine 5′-phosphate	i^6AMP
H	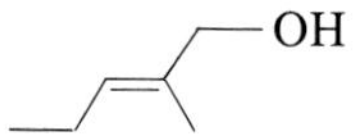	H	Zeatin	io^6Ade
H	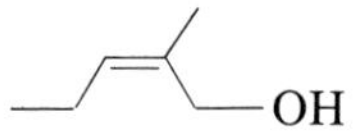	H	cis-Zeatin	c-io^6Ade
H	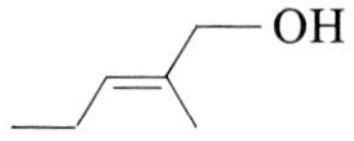	Ribosyl	Zeatin riboside	io^6Ado
H	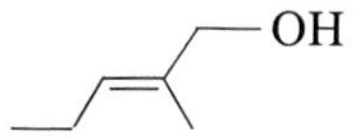	5′-phospho-ribosyl	Zeatin ribotide	io^6AMP
H	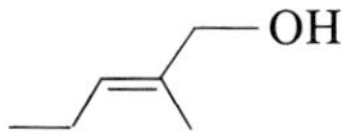	Glucosyl	Zeatin-9-glucoside	io^6Ade-9G
CH_3S	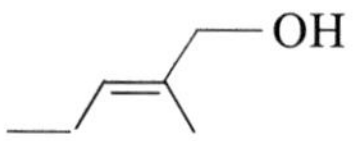	H	2-methylthiozeatin	ms^2io^6Ade
H	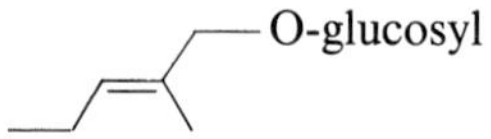	H	Zeatin O-glucoside	io^6Ade OG
H	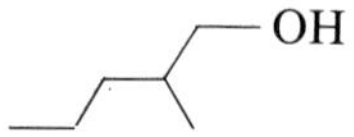	H	Dihydrozeatin	dHio^6Ade
H	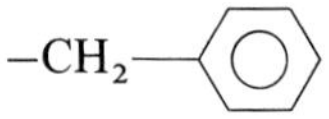	H	Benzyladenine	b^5Ade
H	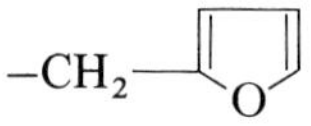	H	Kinetin	f^6Ade
H	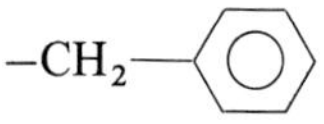	Glucosyl	Benzyladenine-9-glucoside	b^6Ade-9G

chain(s), the sugar, the phosphate and the degree of purine ring and/or side chain modification. Cytokinin bases are both weak acids and bases. For example, the pKa of i^6Ade is 3.4 and the pKb is 10.4 (Leonard and Fujii 1964). Consequently, at pH above 10.4 i^6Ade is negatively charged and at pH below 3.4 it is positively charged. On the other hand, cytokinin ribosides show weak basicity, since the imidazole-NH-groups are blocked by the ribosyl groups (pKa value for i^6Ado is 3.76, Hall 1971). Cytokinin ribotides again are amphoteric because of the phosphate group. Therefore, the retention on a reversed-phase column or on an ion-exchange column can be altered by changes in pH. The structures of some of the cytokinins are listed in Table 1.

2 Cytokinin Isolation and Sample Preparation

The procedures used for cytokinin extraction and sample preparation are usually the important steps in determining the sensitivity and precision of the analysis. Therefore, care must be taken in the execution of the cytokinin isolation and sample preparation.

2.1 Cytokinin Isolation

The qualitative and quantitative aspects of cytokinins extracted from plant tissues vary with the types of extraction solvent and procedure employed. To prevent any enzymatic degradation of cytokinins including conversion of cytokinin ribotide to its riboside (Chen 1981), plant material should be immediately frozen or extracted instantly with proper solvent(s) after harvesting. The isolation and identification of nanomolar quantity of cytokinin (less than 1 nM per gram tissue) are often hampered by the presence of an excess of polyphenols, carbohydrates, terpenoids, and other impurities in plant cells.

The initial step in cytokinin extraction is usually carried out by the homogenization of plant materials with a water-miscible solvent such as ethanol or methanol. After two or three extractions with the same solvent or combinations of various solvents, the combined extract is then subjected to solvent partitioning. If not extracted, plant materials should be either frozen or stored at low temperature.

As shown in Table 2, within the last decade investigators have developed various extraction procedures to isolate cytokinins. These extraction procedures have relied upon a sequence of steps including extraction of the tissue in organic solvents, partition into n-butanol [which has a very high partition coefficient for cytokinin bases (Letham 1974)] or other organic solvents, removal of excess solvent in vacuo, preliminary chromatography on ion-exchange resins or Sephadex LH-20 column followed by HPLC.

Table 2. Examples of solvents used in extraction and partition of cytokinins and related compounds

Tissues	Extraction solvent	Partition solvent	Cytokinins isolated	Ref.
Zea mays (immature kernel)	95% ethanol pH 3.4	n-Butanol and H_2O at pH 7 and 8.2 (together with various column fractionations)	io^{6}AMP io^{6}Ade-9G io^{6}Ado io^{6}Ade other modified io^{6}Ade	Letham (1973)
Vinca rosea crown gall	Cold ethylacetate→ethanol→2-butanol	2-Butanol, H_2O, and ethanol	io^{6}Ade-9G io^{6}AdeOG	Peterson and Miller (1977)
Acer saccharum seed	80% aqueous methanol	Ethylacetate at pH 2.5, followed by H_2O sat.-n-butanol at pH 8	t-io^{6}Ade, c-io^{6}Ade i^{6}Ade, i^{6}Ado	Walker and Dumbroff (1982)
Lupinus luteus developing seed	Methanol-water-acetic acid (70:30:3)→70% methanol	–	io^{6}Ade, io^{6}Ado dHio^{6}Ado, io^{6}AdeOG, io^{6}AdoOG dHio^{6}AdeOG, dHio^{6}AdoOG	Summons et al. (1981)
Nicotiana tabacum cells	(1) methanol:chloroform:formic acid:water (12:3:1:4, v/v) (2) ethylacetate	–	i^{6}Ade, i^{6}Ado i^{6}AMP, i^{6}ADP i^{6}ATP, i^{6}Ade-7G	Laloue and Pethe (1982)
Pisum sativum shoots and *Daucus carota* root tissues	95% ethanol→H_2O sat.-ethylacetate→60% ethanol	–	i^{6}Ade, i^{6}Ado i^{6}Ade, io^{6}Ade i^{6}AMP, io^{6}AMP	Chen et al. (1985)
Vinca rosea crown gall	Chloroform:methanol:formic acid:H_2O (5:12:1:2)→methanol:formic acid:H_2O (6:1:4)→chloroform:methanol:formic acid:H_2O (5:12:1:2)	H_2O-sat.-n-butanol	io^{6}Ade, io^{6}Ado io^{6}AMP	Scott et al. (1982) Palni et al. (1983)

2.2 Sample Preparation

The presence of large amounts of contaminants relative to the material of interest usually results in overloading of the column and poor separations. For this reason, it is wise to precede chromatographic analysis of the sample with the removal of proteins, polysaccharides, and particulates. Organic solvents do not completely remove the protein and polysaccharides, and incomplete removal of these contaminants will shorten the lifetime of the column.

Centrifugation or filtration through a Swinny stainless steel filter (Millipore, HA grade, 0.45 μ filter; for organic solvents, FH grade filter) can be employed to remove macromolecules. The FH filter is less suitable for very small volume samples due to some unavoidable losses in the filter system. The Swinny filter does, however, give an excellent clean-up of the sample.

The solubility of the sample in the organic solvent should be checked before a gradient analysis to ensure that precipitation does not occur later in the gradient run. Therefore, the sample should be dissolved in exactly the same solvent as the mobile phase used in an isocratic analysis, or in the starting phase for a gradient analysis.

3 Preparation of Mobile Phase

3.1 Water

It is essential that all solvents including water be free of impurities; extraneous substances will interfere with UV detection of cytokinins and their derivatives. Impurity may create false optical density peaks. A water purification system, which consists of a combination of two mixed bed ion exchangers, a carbon filter and a 20 μ Millipore filter and an all-glass still can be used to eliminate impurities satisfactorily.

3.2 Organic Solvents

Organic solvents used in HPLC should be HPLC grade solvents. The AR grade solvents have been shown to contain impurities which often give severe problems in gradient analysis. The solvents must be stored in glass and under oxygen-free nitrogen. Organic solvents can be purified according to procedures suggested by Hancock and Sparrow (1984).

Many organic solvents are less stable when mixed with the aqueous component of the mobile phase. For example, acetonitrile is unstable to aqueous acid solution. Therefore, the mobile phases should be prepared freshly each day. The most common organic solvent used in reverse-phase HPLC for cytokinin separation are methanol, acetonitrile, and ethanol, with aqueous phase being either water or buffer with a desired pH value. Ion-pairing reagents have been used in HPLC separation of cytokinins. For example, Walker and Dumbroff (1982) tested N-1-heptanesulphonic acid and 1-octanesulphonic acid as ion-pair reagents and separated several species of cytokinins. It is important to check the solubility of the ion-pairing reagent in several solvent combinations before use. This is to prevent the formation of precipitate which may clog columns and connecting lines.

Before the mobile phases can be used in the chromatograph, each of the phases must be filtered through a 0.45 μ filter to remove insoluble material and to sterilize the solution. It is also necessary to remove dissolved gases in the mobile phase for effective performance of the pumps and optimal separations. There are several ways to remove gases, the most convenient being that of vacuum aspiration.

4 Chromatography

4.1 Column Selection

Various modes of column have been used in the separation and identification of cytokinins. The reversed phase mode is generally used for the analysis of various species of cytokinins; anion-exchange is often the choice for the ionic cytokinin ribotides. More recently, ion-pairing techniques have been developed for the reversed phase separation of cytokinin bases, their ribosides and ribotides. Table 3 gives examples of the types of columns and conditions used by various investigators for cytokinin separations.

There are many factors which dictate the selection of a desirable column for cytokinin or any sample separations. For example, it is important to measure the capacity factor (k′) when a new or different column is used. The rule is that k′ must be greater than 2, a clear-cut boundary which is unaffected by flow rate, column size or other parameters (Hancock and Sparrow 1984).

Isocratic elution is adequate for the HPLC analysis of simple mixtures of synthetic cytokinins. Solvent gradients are necessary for complicated mixtures of extracted cytokinins of widely varying polarity. Thermostatic control of the column is not normally necessary. The most appropriate detector is a spectrophotometer with wavelengths ranging from 254 to 280 nm.

4.2 Separation Examples

4.2.1 Ion Exchange HPLC

In general, anion exchange columns contain quaternary ammonium groups ($-^{+}NR_3$) and cation exchangers contain sulfonic acid groups ($-SO_3^-$). Anion exchange columns have been used more widely than cation exchangers and separation applications have concentrated on cytokinin nucleotides. Laloue and Pethe (1982) achieved the analysis of cytokinin nucleotides (mono-, di-, and tri-phosphates) on an anion exchanger Partisil 10-SAX. The mobile phase contained

Table 3. Examples of HPLC separation and characterization of cytokinins and related compounds

Column/detection	Mobile phase	Compounds separated and identified	Ref.
1. Adsorption column			
Hypersil/265 nm	Ammonia chlorform: methanol: water (100:5:0.3)	i^6Ade, io^6Ade, i^6Ado, ms^2i^6Ado, io^6Ade, Ade, Ado	Horgan and Kramers (1979)
Partisil/265 nm	Tetrahydrofuran: water (9:1)	i^6Ade9G, i^6Ade7G, $dHio^6$Ade	Horgan and Kramers (1979)
2. Normal-phase column LiChrosorb 10 NH_2/254 nm	Hexane and ethanol	i^6Ade, b^6Ade, f^6Ade, i^6Ado, b^6Ado, f^6Ado, c-io^6Ade, c-io^6Ade, t-io^6Ade, c-io^6Ado $dHio^6$Ade	Ernstsen and Jensen (1985)
Partisil 10 PAC/165 nm	Acetonitrile: water (9:1)	io^6Ade-9G, io^6Ade-7G	Horgan and Kramers (1979)
3. Ion-exchange column			
Hypersil APS/265 nm	100 mM triethylammonium bicarbonate in 10 mM ammonium bicarbonate	io^6AMP, i^6AMP, io^6AMPOG	Scott and Horgan (1982)
Partisil 10 SAX	1 mM KH_2PO_4 at pH 5 containing 7% ethanol	i^6AMP, i^6ADP, i^6ATP, AMP, ADP, ATP	Laloue and Pethe (1982)
4. Ion-pair reverse-phase column			
Ultrasphere-IP/269 nm	A 3-solvent linear mobile phase (a) 93% water with 0.2 M acetic acid, pH 2.65 and 6 mM heptane-sulphonic acid (b) 5% acetonitrile with 0.2 M acetic acid (c) 2% methanol with 0.2 M acetic acid programmed to 70% (a), 25% (b), and 5% (c) over 30 min	t-io^6Ade, t-io^6Ado, c-io^6Ade, c-io^6Ado, i^6Ade, i^6Ado	Walker and Dumbroff (1982)
Hypersil-ODS/265 nm	2-segment linear gradient of methanol in 0.1 M acetic acid (with 2.3 mM tetrabutylammonium hydroxide)	i^6AMP, io^6AMP, io^6AMPOG	Scott and Horgan (1982)

Table 3 (continued)

Column/detection	Mobile phase	Compounds separated and identified	Ref.
5. Reverse-phase column			
μ Bondapak C_{18}/280 nm	10 to 30% methanol in 30 min or isocratic elution with 35% methanol	io^{6}Ade, io^{6}Ado, i^{6}Ade, i^{6}Ad, dHio^{6}Ado	Tay et al. (1985)
LiChrosorb (RP-18)/254 nm	Linear gradient of 4% aqueous acetic acid to 100% methanol	i^{6}Ade, i^{6}Ado, io^{6}AMP, io^{6}Ade, io^{6}Ado, io^{6}AMP, Ade, Ado, AMP	Chen et al. (1985)
Hypersil-ODS/265 nm	3-segment linear gradient, water (pH 7) to 11% acetonitrile over 7 min 11 to 20% acetonitrile over 9 min, 20 to 40% acetonitrile over 15 min	io^{6}AdeOG, io^{6}AdoOG, i^{6}Ade, i^{6}Ado, io^{6}Ade-9A, Ade, Ado	Horgan and Kramers (1979)
Octadecyl-silica (Altex ultra-sphere-ODS/254 nm)	Organic phase: 15 to 80% methanol over 30 min. Aqueous phase: 0.02 M acetic acid adjusted to pH 3.5 with ammonia or trimethyl-amine	t-io^{6}Ade, t-io^{6}Ado, c-io^{6}Ade, c-io^{6}Ado, f^{6}Ade, i^{6}Ade, i^{6}Ado, ms^{2}i^{6}Ado	Morris et al. (1981)
μ Bondapak C_{18}/254 nm	40% aqueous methanol	i^{6}Ade 8-OH, i^{6}Ade	Chism et al. (1984)
Ultrasphere-ODS/260 nm	5 to 50% methanol in 0.2 M acetic acid adjusted to pH 3.5 with triethylamine	Various io^{6}Ade metabolites	Lee et al. (1985)
ODS-silica/280 nm	42% methanol 58% water or 20% methanol 80% water	b^{6}Ade b^{6}Ade-7G b^{6}Ade-9G io^{6}Ade, io^{6}Ade-7G	Entsch et al. (1979)

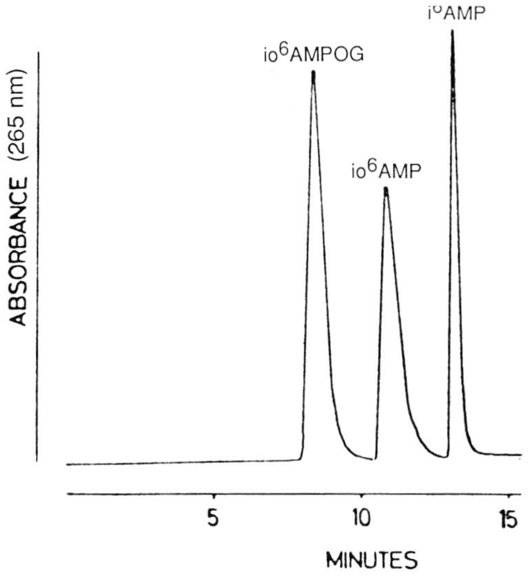

Fig. 1. Separation of cytokinin nucleotides by ion-exchange HPLC. A column (150 × 4.5 mm ID) of Hypersil APS was eluted at a flow-rate of 2 cm^3 min^{-1} with a linear gradient of 100 mM triethylammonium bicarbonate (9–40% over 20 min) in 10 mM ammonium bicarbonate. (Adapted from Scott and Horgan 1982 and reproduced, with permission from Elsevier Scientific Publishing Co)

1 mM KH_2PO_4 at pH 5 and 7% ethanol (solvent A) over 10 min, followed by a multi-step gradient of solvent A to 750 mM KH_2PO_4 at pH 5.5 containing 7% ethanol over 30 min. The authors were able to separate i^6AMP from AMP, i^6ADP from ADP, and i^6ATP from ATP. Scott and Horgan (1982) reported that cytokinin nucleotides could be separated on a column of Hypersil APS, an aminopropyl-bonded silica with weak anion-exchange properties, using volatile buffers. Figure 1 shows the separation achieved using a gradient of increasing proportion of 100 mM triethylammonium bicarbonate in 10 mM ammonium bicarbonate. They stated that complete, although slightly reduced, resolution was also achieved using 1 mM acetic acid in place of ammonium bicarbonate.

4.2.2 Normal Phase HPLC

Traditional normal phase chromatography uses a polar column (e.g., silica gel or alumina) and a nonpolar mobile phase (e.g., hexane or benzene). In general, if the sample is insoluble in aqueous systems, then normal phase chromatography is used. A major advantage of the normal phase procedure is the usage of volatile mobile phases.

Although normal phase has been used for preparative separation of cytokinin bases and their corresponding nucleosides, reverse phase chromatography is preferred for the majority of cytokinin samples since it possesses greater resolving power. Also, if the sample has ionic groups, then normal phase chromatography cannot be used because of low solubility of the sample in nonaqueous mobile phases. In addition, poor peak shapes and separations are often observed when silica columns are used with polar samples.

Using hexane and ethanol as mobile phase, Ernstsen and Jensen (1985) were able to separate all the cytokinin ribosides with the exception of i^6Ado and b^6Ado. The column used was a 250 × 4.6 mm ID LiChrosorb 10 NH_2 (Fig. 2).

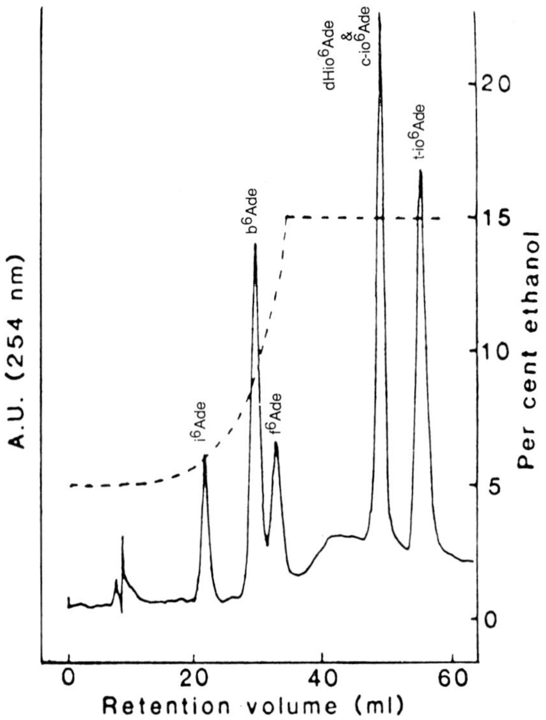

Fig. 2. Normal-phase liquid chromatography of cytokinins. Gradient elution of free bases with ethanol (10 mM in ammonium acetate) and n-hexane. Column: LiChrosorb 10 NH_2. Concave gradient: 5% to 15% ethanol in 15 min. Flow-rate: 2 ml min^{-1}. (After Ernstsen and Jensen 1985)

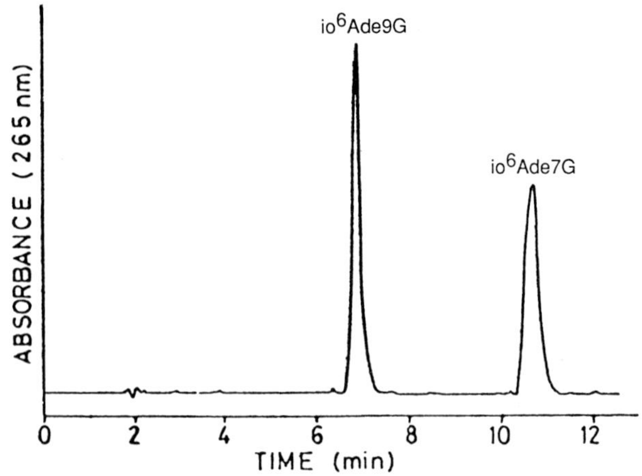

Fig. 3. Separation of isomeric io^{6}Ade N-glucosides by normal-phase partition HPLC. Column: Partisil 10 PAC (250 × 4.5 mm ID). Flow-rate: 2 ml min^{-1}. Mobile phase: acetonitrile-water (9:1). (After Horgan and Kramers 1979)

Thus, normal-phase HPLC may provide a tool for partly separating cytokinin ribosides as a group from cytokinin bases, and this system is not advisable to use in the final, quantitative step. Horgan and Kramers (1979) investigated the use of Partisil 10 PAC, a normal-phase partition column, to separate polar cytokinins. Cytokinin glucosides such as io^{6}Ade-7G and io^{6}Ade-9G were successfully separated on this column using a solvent system of acetonitrile: water (9:1, v/v) (Fig. 3).

4.2.3 Reverse Phase HPLC

In reverse phase chromatography the sample molecules react preferentially with either the polar mobile phase or the nonpolar support. There are wide ranges of functional groups available as chemically bonded stationary phase. Nevertheless, all column packings function by providing a nonpolar surface area for absorption of the solute from the polar mobile phase. Each support has specific application, and thus there are many parameters which determine the choice of a column suited for a given separation. For example, particle size, surface area of the silica particles per gram, amount of reversed phase per gram, the presence of polymer-

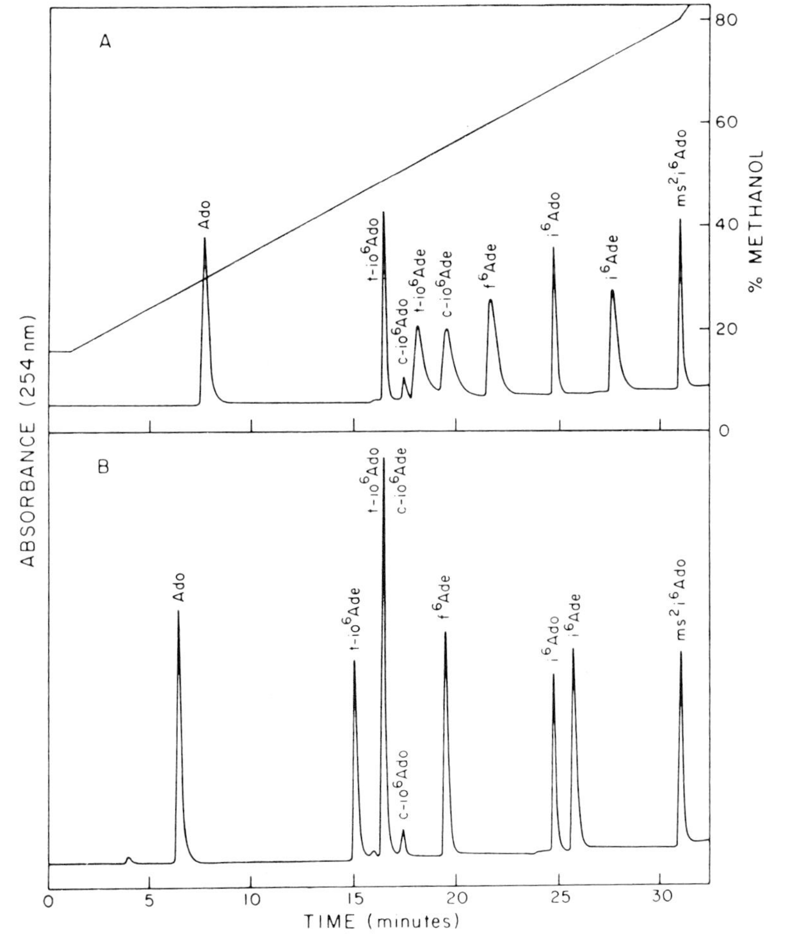

Fig. 4. A, B. High-performance liquid chromatography of standard cytokinins. Column: octadecyl-silica (Altex Ultrasphere ODS, 5 μ, 25 cm × 4.6 mm). Load: 100 ng each component. Buffer: 0.02 M acetic acid adjusted to pH 3.5 with **A** ammonia, **B** triethylamine. Flow rate: 1.5 ml min^{-1}. Gradient: 15% to 80% methanol over 30 min. Detection: absorbance at 254 nm. (After Morris et al. 1981)

ized surface coating and stability of the reversed phase to aqueous conditions are all important column parameters (Hancock and Sparrow 1984).

The reversed phase mode is the most common method for separation of cytokinin bases and corresponding nucleosides. The columns are normally very efficient and stable. A large number of applications of cytokinin separations using reversed phase columns have been reviewed (Koshimizu and Iwamura 1986).

For separation of various cytokinin bases and nucleosides, Morris et al. (1981) employed microparticulate octadecyl silica columns (Altex Ultrasphere-ODS) (Fig. 4). The cis- and trans-io^{6}Ade and corresponding ribosides are readily resolved from each other in a pH 3.0 acetate buffer by elution of the column with linearly increasing concentrations of acetonitrile, ethanol or methanol. Analysis time is short and column capacity is high. Tailing of the free bases may be eliminated by the addition of suitably competitive substances such as imidazole or triethylamine to mask free residual silanols of the matrix. Horgan and Kramers (1979) have systematically investigated various modes of HPLC for cytokinin separations. As shown in Fig. 5, a series of cytokinins related to io^{6}Ade and a number of common purines and pyrimidines were successfully separated by a reverse phase column, Hypersil-ODS.

Reversed phase HPLC has been employed to separate and identify cytokinin metabolites. As shown in Fig. 6, cytokinin glucosides formed by an enzyme were clearly separated from their substrates b^{6}Ade and io^{6}Ade on ODS-silica column (Entsch et al. 1979). Using a LiChrosorb column together with other analytical techniques, Chen et al. (1985) localized six species of cytokinins biosynthesized by pea stems and carrot root cambium. Octadecyl silane (ODS or C_{18}) type of reversed phase columns are the most commonly used materials today to separate and identify cytokinins and related compounds (Scott et al. 1982; Summons et al.

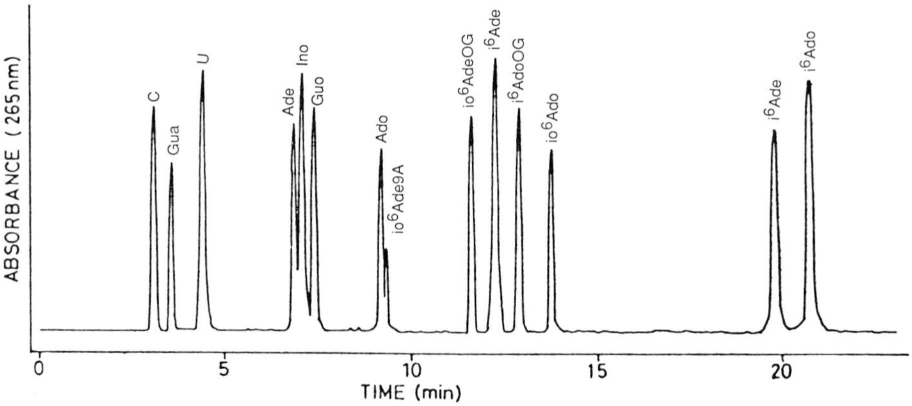

Fig. 5. Separation of a series of cytokinins related to io^{6}Ade and a number of common purines and pyrimidines by reversed-phase HPLC. Column: Hypersil ODS (150 × 4.5 mm ID). Flow rate: 2 ml min^{-1}. Mobile phase: 3-segment linear gradient, water (pH 7 with triethylammonium bicarbonate) to 11% acetonitrile over 7 min, 11 to 20% acetonitrile over 9 min, and 20 to 40% acetonitrile over 15 min io^{6}Ade9A : 1-alanyl io^{6}Ade. (After Horgan and Kramers 1979)

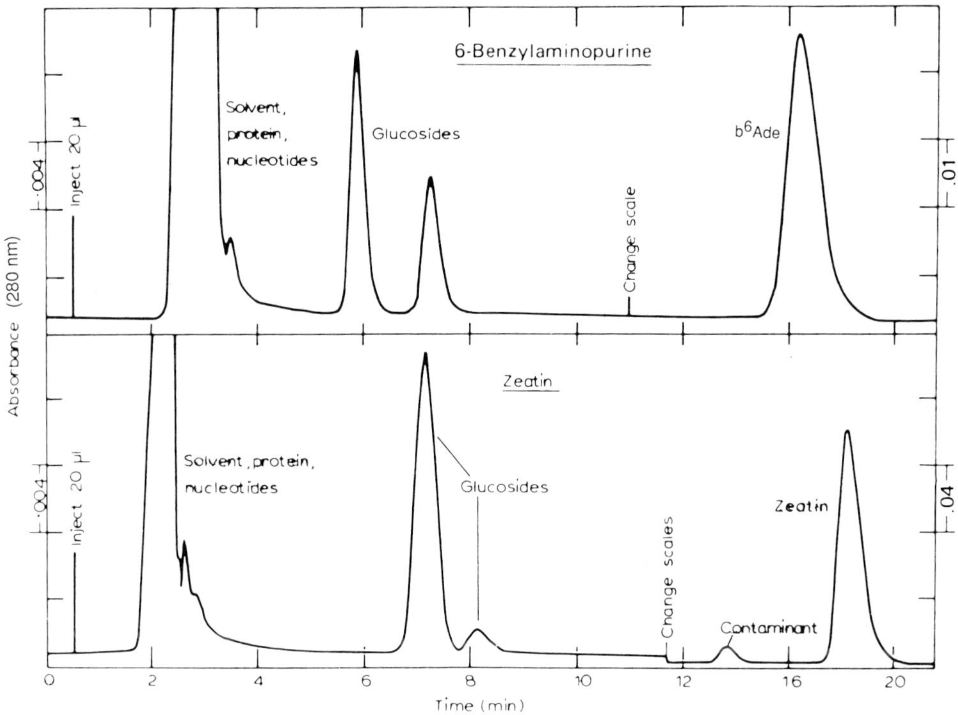

Fig. 6. Quantitative analysis of cytokinin metabolites by reversed-phase HPLC during enzyme assays. The traces were recorded at 280 nm, and resulted from injections of 20 µ aliquots of the respective incubation mixtures. Column: ODS-silica. For benzylaminopurine, the mobile phase was 42% methanol and 58% H_2O, run at 1.0 ml min^{-1}. For io^{6}Ade (zeatin), the mobile phase was 20% methanol and 80% H_2O, run at 1.5 ml min^{-1}. At the position marked "changed scale", the recorder speed was halved for io^{6}Ade only. (After Entsch et al. 1979)

1980; Stahly and Buchanan 1982; Andersen and Kemp 1979; MacDonald et al. 1981; Walker and Dumbroff 1982). Preparative C_{18} reverse phase columns have also been employed in initial purification of cytokinins (Summons et al. 1979; Kannangara et al. 1978).

4.2.4 Ion-Pair Reverse Phase HPLC

The concept of ion-pairing refers to the ionic association between ions added to the mobile phase and oppositely charged ionized groups on the sample molecule. Such an association can lead to a complex of either increased or decreased polarity. As shown in Fig. 7 (Scott and Horgan 1982) and 8 (Walker and Dumbroff 1982), cytokinin nucleotides, along with their corresponding nucleosides and bases, have been separated by ion-pair reverse phase HPLC. The reagents and the conditions were chosen so that the ionic cytokinin nucleotides, which usually elute in the void volume in a reverse phase system, are retained longer on the column without interfering with cytokinin nucleosides and bases which have longer retention times.

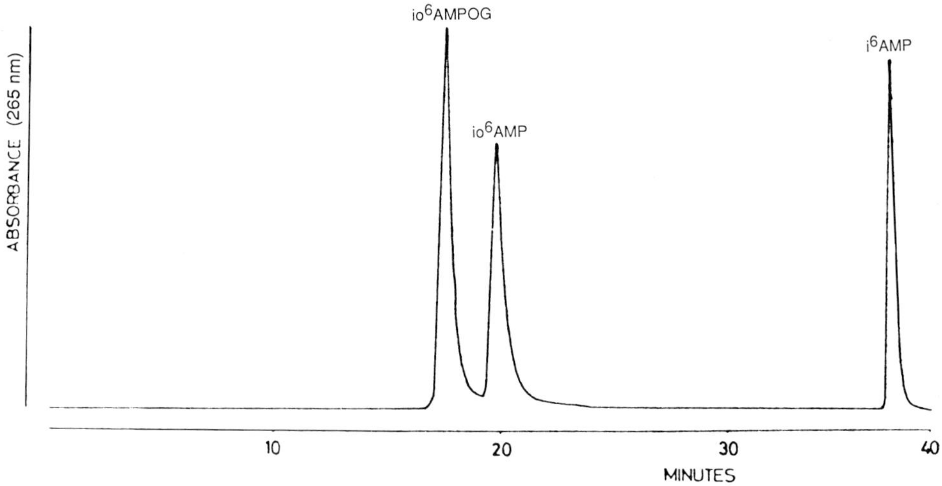

Fig. 7. Separation of cytokinin nucleotides by ion-pair reversed-phase HPLC. A column (150 × 4.5 mm ID) of Hypersil ODS was eluted at a flow-rate of 2 ml/min with a 2-segment linear gradient of methanol in 0.1 M acetic acid containing 2.5 mM tetrabutylammonium hydroxide. Segment 1: 15% to 30% methanol over 30 min. Segment 2: 30% to 90% methanol over 10 min. (After Scott and Horgan 1982)

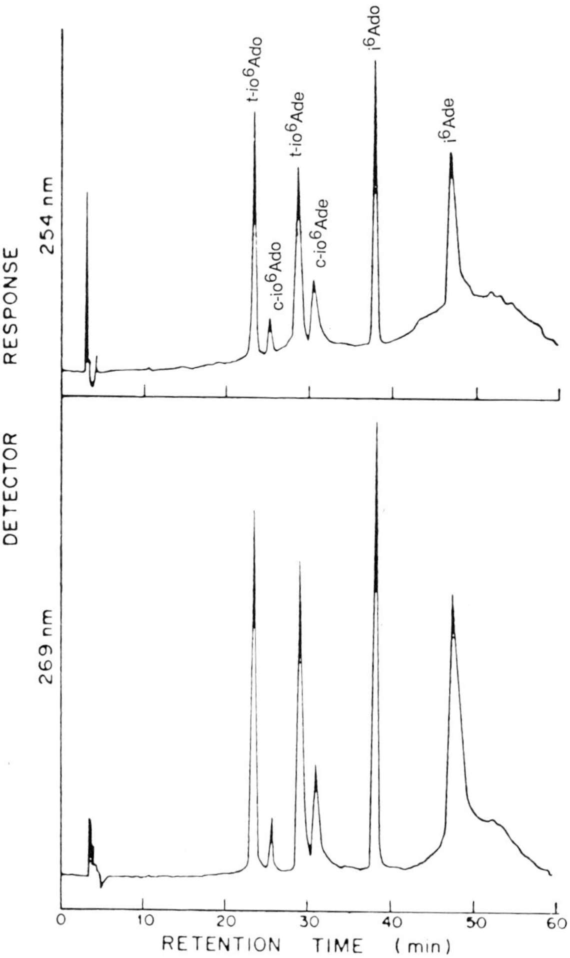

Fig. 8. Separation of cytokinin standard by ion-pair chromatography. Column: Ultrasphere-IP 5 µm (250 × 4.6 mm ID). Flow rate: 0.8 ml min^{-1}. Temperature: 40 °C. A three-solvent mobile phase with a two-segment linear gradient: (A) 93% water with 0.2 M acetic acid, pH 2.65, and 6 mM heptanesulphonic acid; (B) 5% acetonitrile with 0.2 M acetic acid; (C) 2% methanol with 0.2 M acetic acid programmed to 70% A, 25% B, 5% C over 30 min, to 50% A, 45% B, 5% C over 8 min and held for 7 min. (After Walker and Dumbroff 1982)

5 Concluding Remarks

HPLC is a powerful tool for use in plant hormone research. From a review of literature, it is clear that virtually all the cytokinins, their metabolites and other related compounds, can be separated and analyzed using the combination of HPLC and other chromatographic methods, particularly Sephadex LH-20 column. The HPLC separations are far more rapid than classical methods and provide high resolution and sensitivity. Samples which require final confirmation of their chemical structures usually need the aid of spectroscopic methods such as nuclear magnetic resonance and mass spectroscopy. Cytokinins and related compounds purified by HPLC are often pure enough for direct immunoassay or mass spectrometer analysis.

Column switching technique (reviewed by Hulpke and Werthmann 1985), which has been developed in recent years and has yet to be applied to cytokinin separation, may be a useful technique for cytokinin analysis. It is very likely that future improvements in instrumentation, computerization, and operating conditions will improve significantly the resolution, sensitivity, and time of analysis of cytokinins. Consequently, one is inclined to predict an increase in the importance of HPLC analysis of plant hormones.

Acknowledgments. The research reported was supported in part by NSF DCB-8609777 and USDA Competitive Grant GAM 8502373.

References

Anderson RA, Kemp JR (1979) Reversed-phase high-performance liquid chromatography of several plant cell division factors (cytokinins) and their cis and trans isomers. J Chromatogr 172:509–512

Brenner ML (1981) Modern methods for plant growth substance analysis. Annu Rev Plant Physiol 32:511–538

Chen CM (1981) Biosynthesis and enzymic regulation of the interconversion of cytokinin. In: Guern J, Péaud-Lenoël C (eds) Metabolism and molecular activities of cytokinins. Springer, Berlin Heidelberg New York, pp 34–43

Chen CM, Ertl JR, Leisner SM, Chang CC (1985) Localization of cytokinin biosynthetic sites in pea plants and carrot roots. Plant Physiol (Bethesda) 78:510–513

Chism GW III, Long AR, Rolle R (1984) Assay of enzymes involved in cytokinin metabolism by means of reversed-phase high-performance liquid chromatography. J Chromatogr 317:263–270

Engelhardt H (ed) (1985) Practice of high performance liquid chromatography: Applications, equipment, and quantitative analysis. Springer, Berlin Heidelberg New York, Tokyo

Entsch B, Letham DS, Parker CW, Summons RE (1979) Preparation and characterization, using high-performance liquid chromatography, of an enzyme forming glucosides of cytokinins. Biochim Biophys Acta 570:124–139

Entsch B, Parker CW, Letham DS (1983) An enzyme from lupin seeds forming alanine derivatives of cytokinins. Phytochemistry (Oxf) 22:375–382

Ernstsen A, Jensen E (1985) High performance liquid chromatography of cytokinins. J Liquid Chromatogr 8:369–379

Hall RH (1971) The modified nucleosides in nucleic acids. Columbia University Press, New York, p 102

Hancock WS, Sparrow JT (1984) HPLC analysis of biological compounds. A laboratory guide. Dekker, New York Basel

Horgan R, Kramers MR (1979) High-performance liquid chromatography of cytokinins. J Chromatogr 173:263–270

Koshimizu K, Iwamura H (1986) Cytokinins. In: Takahashi N (ed) Chemistry of plant hormones. CRC Press, Boca Raton, pp 153–199

Laloue M, Pethe C (1982) Dynamics of cytokinin metabolism in tobacco cells. In: Wareing PF (ed) Plant growth substances 1982. Proc 11th Int Conf Plant Growth Subst. Academic Press, New York, pp 185–195

Lee YH, Mok MC, Mok DWS, Griffin DA, Shaw G (1985) Cytokinin metabolism in *Phaseolus* embryos. Genetic difference and the occurrence of novel zeatin metabolites. Plant Physiol (Bethesda) 77:635–641

Leonard NJ, Fujii T (1964) The synthesis of compounds possessing kinetin activity. The use of a blocking group at the 9-position of adenine for the synthesis of 1-substituted adenine. Proc Natl Acad Sci USA 51:73–75

Letham DS (1973) Cytokinins from *Zea mays*. Phytochemistry (Oxf) 12:2445–2455

Letham DS (1974) Regulators of cell division in plant tissues XXI. Distribution coefficients for cytokinins. Planta 118:361–364

Letham DS, Palni LMS (1983) The biosynthesis and metabolism of cytokinins. Annu Rev Plant Physiol (Bethesda) 34:163–197

MacDonald EMS, Akiyoshi DE, Morris RO (1981) Combined high-performance liquid chromatography radioimmunoassay for cytokinins. J Chromatogr 214:101–109

McGaw BA, Scott IM, Horgan R (1984) Cytokinin biosynthesis and metabolism. In: Crozier A, Hillman JR (eds) The biosynthesis and metabolism of plant hormones. Cambridge University Press, Cambridge, pp 105–133

Morris RO, Regier DA, MacDonald EMS (1981) Analytical procedures of cytokinins: Application to *Agrobacterium tumefaciens*. In: Guern J, Péaud-Lenoël C (eds) Metabolism and molecular activities of cytokinins. Springer, Berlin Heidelberg New York, pp 3–16

Palni LMS, Horgan R, Darrall NM, Stuckburg T, Wareing PF (1983) Cytokinin biosynthesis in crown gall tissue of *Vinca rosea*. Planta 159:50–59

Peterson JB, Miller CO (1977) Glucosylzeatin and glucosylribosylzeatin from *Vinca rosea* L. crown gall tumor tissue. Plant Physiol (Bethesda) 59:1026–1028

Scott IM, Horgan R (1982) High-performance liquid chromatography of cytokinin ribonucleoside 5′-monophosphates. J Chromatogr 237:311–315

Stahly EA, Buchanan DA (1982) High-performance liquid chromatographic procedure for separation and quantification of zeatin and zeatin riboside from pears, peaches and apples. J Chromatogr 235:453–457

Summons RE, Entsch B, Letham DS, Gollnow BI, MacLeod JK (1980) Regulators of cell division in plant tissues. XXVIII. Metabolism of zeatin in sweet-corn kernels: purification and identification using high-performance liquid chromatography and chemical ionization mass spectrometry. Planta 147:422–434

Summons RE, Letham DS, Gollnow BI, Parker CW, Entsch B, Johnson LP, MacLeod JK, Rolfe BG (1981) Cytokinin translocation and metabolism in species of the leguminoseae: Studies in relation to shoot and nodule development. In: Guern J, Péaud-Lenoël C (eds) Metabolism and molecular activities of cytokinins. Springer, Berlin Heidelberg New York, pp 69–79

Tay SAB, MacLeod JK, Palni LMS, Letham DS (1985) Detection of cytokinins in a seaweed extract. Phytochemistry (Oxf) 24:2611–2614

Walker MA, Dumbroff EB (1982) Use of ion-pair, reversed-phase, high-performance liquid chromatography for the analysis of cytokinins. J Chromatogr 237:316–323

Separation and Purification of Abscisic Acid and Its Catabolites by High Performance Liquid Chromatography

R. A. CREELMAN and J. A. D. ZEEVAART

Abbreviations

ABA-GE	β-D-glucopyranosyl abscisate
ABA-GS	1′-*O*-abscisic acid-β-D-glucopyranoside
DPA	dihydrophaseic acid
DPA-GS	4′-*O*-dihydrophaseic acid-β-D-glucopyranoside
EI	electron impact
EIA	enzyme immunoassay
GLC-ECD	gas liquid chromatography-electron capture detection
GC-MS	gas chromatography-mass spectroscopy
8′-OH-ABA	8′-hydroxy ABA
HMG-ABA	β-hydroxy-β-methylglutarylhydroxy abscisic acid
HPLC	high performance liquid chromatography
min	minute(s)
Me-ABA	methyl ester of abscisic acid
NCI	negative chemical ionization
ORD	optical rotary dispersion
PA	phaseic acid
PA-GE	β-D-glucopyranosyl phaseate
RIA	radioimmunoassay
SIM	selected ion monitoring
TLC	thin layer chromatography
UV	ultraviolet

1 Introduction

Numerous theories on the role of ABA as a plant growth regulator have been proposed, including, but not limited to, root gravitropism (Wilkins 1978, 1984; but see Moore and Evans 1986), embryo and bud dormancy (Addicott and Lyon 1969; Walton 1980; Wareing 1978), and stomatal closure (Raschke 1975). However, Wareing (1978) and Walton (1980) have concluded that too little is known about the events occurring during any physiological process to state unequivocally that ABA is involved in its control. The best evidence, although still correlative, for the involvement of ABA in a physiological process has been obtained with experiments on water stress. Detached or attached leaves will accumulate ABA (usually 10 to 40 times that found in a turgid leaf) upon the imposition of water stress. When ABA is given to epidermal strips or fed to intact leaves through the transpiration stream, stomata close rapidly (Raschke 1975; Davies and Mansfield 1983). The effect of ABA on stomata is rather rapid compared to its effect on protein synthesis, as in the control of α-amylase production in barley aleurone layers. In this tissue ABA inhibits the gibberellin-induced synthesis of α-amylase, which is needed for the hydrolysis of endosperm starch during germination (Ho 1983, Jacobsen 1983).

ABA has been proposed to be an inhibitor of vivipary (precocious germination) and a promoter of embryo maturation. Vivipary occurs normally in mangroves (*Rhizophora mangle*) and it has been shown that embryos of this species are insensitive to ABA (Sussex 1975). The viviparous mutants of maize *(vp2, vp5, vp9,* and *w3)* have decreased levels of ABA (Brenner et al. 1977; Smith et al. 1978; Moore and Smith 1985). In addition, ABA is required for induction of dormancy in an ABA-deficient mutant of *Arabidopsis thaliana* (Karssen and Laçka 1986).

In other studies, it has been shown that ABA application to excised dicot embryos will inhibit both germination and the appearance of germination-specific proteins. Furthermore, ABA promotes the synthesis of embryo-specific mRNA's and proteins (Ihle and Dure 1970; Dure et al. 1983; Crouch and Sussex 1981; Crouch et al. 1985). The accumulation of wheat germ agglutinin (a lectin found in wheat embryos, shoots, and roots) (Raikhel et al. 1986) and an "early-methionine-labeled" (E_m) polypeptide in cultured immature wheat embryos (Williamson et al. 1985) is induced by ABA.

To gain a better understanding of the physiology of ABA action, it is important to know how the levels of ABA and its catabolites change in the particular organ or tissue that is being investigated. Thus, it is important to measure ABA (and catabolites) in guard cells rather than leaves, and in embryos rather than in intact seeds.

The extent to which a plant extract must be purified for the measurement of ABA and catabolites depends on a number of factors, including the analytical procedure used, the amount of ABA (and catabolites) present, and the nature and amount of interfering compounds found in the extract. Initially, the quantification of ABA was performed by assaying the biological activity of extracts in different bioassays [for example, inhibition of wheat coleoptile elongation, *Lemna* growth inhibition, and stomatal closure (Dörffling and Tietz 1983; Zeevaart 1979)]. However, these assays are of limited use because they are (a) time-consuming, (b) semi-quantitative, and (c) they may have potential inhibition from interfering compounds.

Since its discovery, several methods for the purification and quantification of ABA (and its catabolites) have been developed based on its physical and chemical properties (Table 1). Of these procedures, the most convenient method for the purification of ABA is HPLC. The purpose of this chapter is to describe how HPLC can be used to characterize ABA and its catabolites.

HPLC provides an excellent means by which substantial purification of a crude plant extract may be obtained in a short time, eliminating the need for solvent partitioning and subsequent TLC. Furthermore, HPLC is a useful separation technique when the sample has to be analyzed further, for example, quantification by GLC-ECD, GC-MS, EIA, or RIA. The levels of ABA and its catabolites may be analyzed by HPLC with quantification by UV detection; however,

Table 1. Advantages and disadvantages of various methods for the detection and measurement of abscisic acid and catabolites

Method	Advantages	Disadvantages
ORD	High selectivity	Low sensitivity
GC-FID		Low selectivity and sensitivity; requires derivatization
GC-ECD	High sensitivity and selectivity; small tissue samples needed	Requires derivatization
GC-MS		
EI	High selectivity	Expensive equipment
SIM	High sensitivity	Expensive equipment; May not be selective if a contaminant has similar masses
NCI	High sensitivity and selectivity	Expensive equipment
HPLC-UV	Rapid purification	Not selective or sensitive; must be confirmed by another method
EIA	Rapid; more sensitive than RIA; specific for ABA	Samples need purification by TLC or HPLC; no antibodies for catabolites available; expensive to develop or purchase
RIA	Rapid; more precise than EIA; specific for ABA	Recovery with internal radioactivity standards not possible; samples need purification by TLC or HPLC; no antibodies available for catabolites; expensive and time-consuming to develop or purchase

this technique has the disadvantages of lower selectivity and sensitivity compared with GLC-ECD or RIA. If samples are quantified by HPLC (or any immunological method; see Wang et al. 1986), confirmation of identity must be obtained by GLC-ECD or GC-MS. Analysis of ABA and other plant growth substances by HPLC and other physicochemical methods has recently been reviewed (Brenner 1981).

2 Abscisic Acid and Its Catabolites

ABA (molecular weight 264) is a sesquiterpenoid and contains a carboxyl, keto, and hydroxyl group (Fig. 1). It is a weak acid (pK_a 4.8) and partitions into organic solvents, such as diethyl ether, dichloromethane, and ethyl acetate (but not hexane), at low pH. The molecule has one chiral center at C-1′. The naturally occurring enantiomer is dextrorotary and has the sinister (S) configuration. Synthetic material is a racemic mixture of the two enantiomers. Thus, any results describing the metabolism of ABA should describe the type of ABA used because the metabolism of (+)- and (−)-ABA is different (see below). The naturally oc-

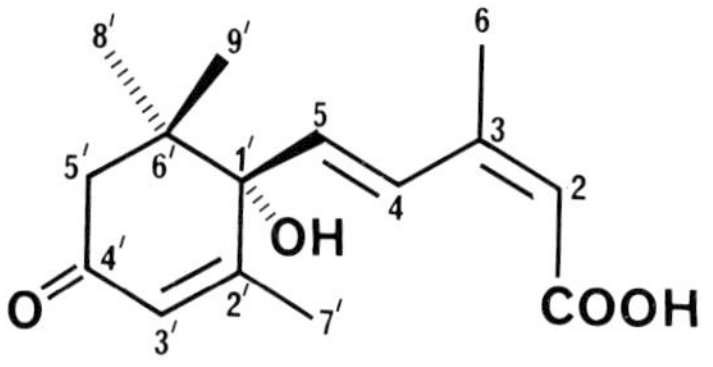

Fig. 1. Structure of abscisic acid. The numbering system in the revised one described in Boyer et al. (1986)

curring enantiomer, (+)-ABA, shows extremely high optical activity, with extrema at 289 nm ($\alpha = +24{,}000°$) and 246 nm ($\alpha = -69{,}000°$). This property has been used to quantify ABA by ORD. However, this method requires extensive purification of a fairly large quantity of ABA (Milborrow 1967). ABA absorbs in the UV, with its maximum varying with the pH of the solution (Dörffling and Tietz 1983). In acidic ethanol, the maximum is at 260 nm ($\varepsilon = 21{,}400$; Milborrow 1967). In addition to optical isomers, geometric isomers are also present. The configuration of the side chain at C-2 can be either cis or trans. By convention, naturally occurring 2-cis,4-trans-ABA is called ABA, and the 2-trans,4-trans isomer *t*-ABA (Addicott et al. 1968). Strong light catalyzes the isomerization at the C-2, C-3 double bond to establish an approximate 1:1 ratio of ABA:*t*-ABA in solution.

ABA is rapidly catabolized in plants by either conjugation to water-soluble catabolites or by oxidation to more polar compounds (Fig. 2). This section will deal with catabolites that have been fully characterized. As the catabolism of ABA in different plant species is investigated, additional catabolites will undoubtedly be detected and characterized (Loveys and Milborrow 1984; Walton 1980).

A major pathway in higher plants involves the hydroxylation of ABA at the 8′-geminal methyl group to give the unstable intermediate 8′-OH-ABA, which rearranges to form PA, and the subsequent reduction to DPA. This pathway has been shown to operate in several plants, including beans (Harrison and Walton 1975), ash seeds (Sondheimer et al. 1974), and castor bean (Zeevaart 1977). In spite of many attempts to isolate it, 8′-OH-ABA has only been reported once (Milborrow 1969). Acetylation of short-term incubation mixtures in a cell-free system gave evidence for the existence of 8′-OH-ABA as an intermediate in PA biosynthesis (Gillard and Walton 1976). The epimer of DPA has also been reported, but the content was only 1 to 2% of that of DPA (Zeevaart and Milborrow 1976). ABA is also reduced to the 1′,4′-trans-diol (Milborrow 1983; Vaughan 1986) and 1′,4′-cis-diol (Dathe and Sembdner 1982). The diols are unstable in acid (Vaughan 1986) and are endogenous compounds (Dathe and Sembdner 1982; Vaughan 1986).

In addition to catabolism to PA and DPA, conjugation of ABA also occurs. ABA-GE is widely distributed in plants, but has been unequivocally identified only a few times (Hirai et al. 1978; Boyer and Zeevaart 1982b; Neill et al. 1983). However, ABA-GE is not the sole constituent of base-labile conjugates of ABA. For instance, Loveys and Milborrow (1984) reported the occurrence of another ABA conjugate, ABA-GS, in tomato. Other conjugates of ABA catabolites, such as HMG-ABA (Hirai et al. 1978), PA-GE (Boyer and Zeevaart 1982a) and DPA-

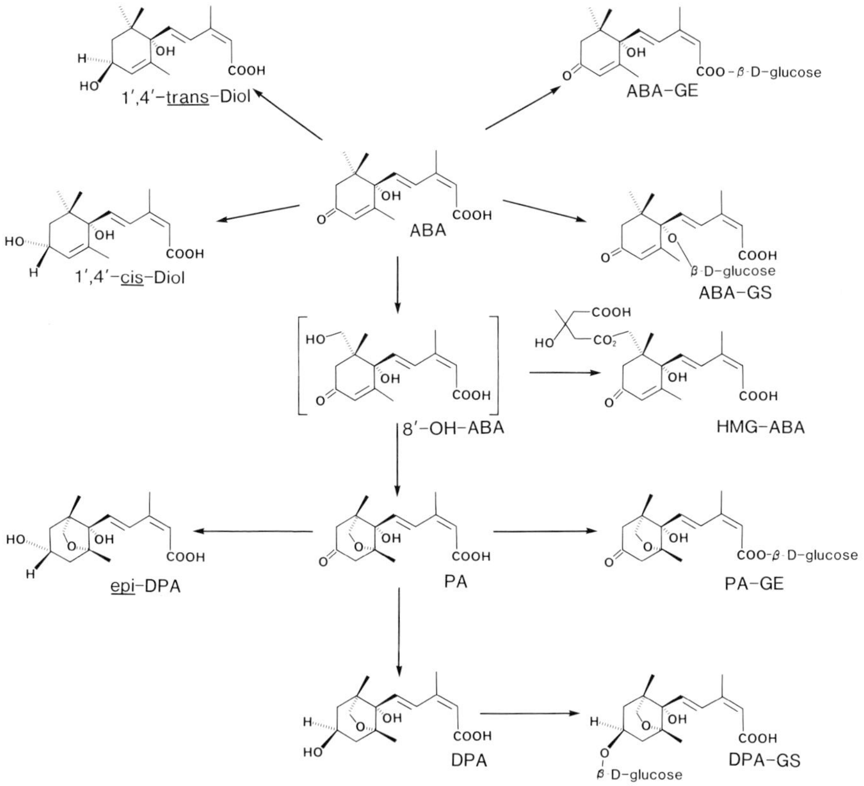

Fig. 2. Catabolites of abscisic acid. The catabolites shown are those which have been unequivocally identified, and are not meant to indicate that all compounds are found in any one plant

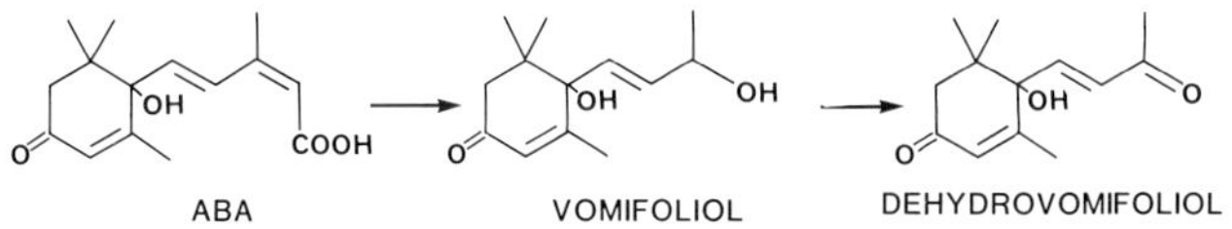

Fig. 3. Catabolism of abscisic acid in *Corynebacterium* according to Hasegawa et al. (1984)

GS (Milborrow and Vaughan 1982; Hirai and Koshimizu 1983) have also been described.

Cleavage of the side chain of ABA at C-2, C-3 to form vomifoliol may occur. However, this is based on structural similarity and in no case has the conversion of ABA to vomifoliol ever been shown in higher plants. Conversion of ABA to dehydrovomifoliol and the presence of vomifoliol dehydrogenase activity have been demonstrated in a species of *Corynebacterium* (Fig. 3; Hasegawa et al. 1984).

3 Use of High Performance Liquid Chromatography to Characterize Abscisic Acid and Its Catabolites

3.1 Synthesis of (±)-Abscisic Acid and Catabolites

There are several commercial sources for ABA, including both ^{14}C- and ^{3}H-labeled analogs. The synthesis of ABA starts with the oxidation of α-ionone with *t*-butyl chromate to give (±)-6-hydroxy-3-keto-α-ionone (Roberts et al. 1968). After a Wittig reaction with (carbethoxymethylene)triphenylphosphorane and basic hydrolysis with methanolic potassium hydroxide a mixture of ABA and *t*-ABA is formed which may be separated by crystallization from ether or by chromatography. Radioactivity is introduced into the ABA molecule at two possible steps. (±)-[2-^{14}C]-ABA is synthesized by using methyl [2-^{14}C]bromoacetate to make either trimethyl-[2-^{14}]phosphonacetate for use in a Horner reaction (Cornforth et al. 1968) or (carbmethoxy-^{14}C-methylene)triphenylphosphorane in a Wittig reaction (Lehmann et al. 1973). ^{3}H is introduced (primarily into the side chain methyl group of ABA) by subjecting (±)-6-hydroxy-3-keto-α-ionone to a keto-enol exchange reaction in $^{3}H_2O$ (Walton et al. 1977). The tritium can no longer exchange from the side chain methyl group after the subsequent Wittig reaction.

There are no commercial sources for the catabolites of ABA (either unlabeled or radioactive). Standards must either be obtained by extraction of plant material (Zeevaart and Milborrow 1976), or through synthetic procedures (for ABA-GE see Lehmann et al. 1975; for the methyl ester of PA see Takahashi et al. 1986).

3.2 Resolution of (±)-Abscisic Acid

Recently, two methods have been reported for the resolution of the enantiomers of ABA. Vaughan and Milborrow (1984) resolved the enantiomers by first methylating (±)-ABA and then reducing Me-ABA with $NaBH_4$ to give methyl (±)-1′,4′-trans- and (±)-1′,4′-cis-diol of ABA. The products were separated to obtain pure methyl (±)-1′,4′-cis-diol-ABA (isocratic, ethanol:water:acetic acid 317:517:1, v:v:v, flow rate 4.0 ml min^{-1}, 30 × 0.78 cm column, μBondapak C_{18}, Waters Associates, Milford, MA, USA). It was found that only (±)-methyl 1′,4′-cis ABA diol could be resolved further. The best resolution of the optical isomers of (±)-methyl 1′,4′-cis diol of ABA was achieved by connecting a Pirkle Type 1-A column (Regis, Morton Grove, IL, USA) and a Pirkle covalent *R*-phenylglycine column (Regis) in series and recycling the eluate (hexane: *iso*-propanol 97:3, v:v, flow rate 2.0 ml min^{-1}) four times [the (+) enantiomer eluted ahead of the (−) enantiomer]. ABA was regenerated by treating the methyl diol with active MnO_2 followed by alkaline hydrolysis. Vaughan and Milborrow (1984) mention that the length of time (248 min for 300 μg) was compensated for by obtaining complete baseline resolution of the diols.

Another method was described utilizing monoclonal antibodies [produced against (+)-ABA] bound to a Sepharose column (Knox and Galfre 1986). Because of the low capacity (20 μg could be resolved within 4 h), the use of a monoclonal antibody column is primarily for the resolution of radioactive (±)-ABA rather than unlabeled synthetic ABA.

If large quantities of (+)-ABA are needed, the material may be isolated from the fungus *Cercospora rosicola* (Assante et al. 1977), or from other natural sources, such as red wine (Loveys and Downton 1979), or cider apple juice (Gaskin and MacMillan 1968). However, other than by the resolution of the optical isomers of synthetic ABA, there is no method for obtaining the unnatural enantiomer (−)-ABA.

3.3 Methods for the Purification of Abscisic Acid and Catabolites by High Performance Liquid Chromatography

There is no single, universal method for the extraction of ABA and catabolites from plant material. Most procedures involve homogenization and extraction in acidic acetone containing an antioxidant (for instance, 1% acetic acid, 100 mg l^{-1} butylated hydroxytoluene [2,6-di-*tert*-butyl-4-methylphenol]; Cornish and Zeevaart 1984). Methanol may be used as the extraction solvent, but must be acidified to prevent the transmethylation of ABA-GE to Me-ABA (Milborrow and Mallaby 1975). After extraction and removal of the organic solvent (by rotary evaporation), the aqueous residue may be frozen and lyophilized. The dry extract is then dissolved in the initial solvent used for HPLC analysis and filtered through a 0.45 μm filter to remove insoluble material.

Initial purification of ABA and its catabolites is usually performed with a C_{18} reverse phase column (Fig. 4). If the initial solvent is 1% aqueous acetic acid or a low percentage organic solvent composition, ABA and catabolites are retained at the head of the column. Hence, large injection volumes (2–5 ml) may be used. This last point is essential for the purification of ABA, since it is not practical to dissolve a crude plant extract in volumes of 0.1 to 0.5 ml. The addition of a water-miscible organic solvent (such as ethanol, methanol, acetonitrile, or a mixture of these three) causes the separation and elution of ABA and catabolites from the column. The order of elution depends on the polarity of the compounds, with the most polar catabolite eluting first. The degree of separation depends on the steepness of the gradient and the initial and final solvent compositions. After many runs (usually >1000), column degradation will be noticed as a change in the retention time of standards. If this occurs, increasing the initial solvent polarity will usually rectify the problem.

A major separation problem is the resolution of ABA-GE and PA by HPLC (Fig. 4). For the analysis of ABA, PA, and ABA-GE, Cornish and Zeevaart (1984) used a linear gradient from 10% to 35% ethanol in 1% acetic acid in 40 min (flow rate 2.5 ml min^{-1}) with a μBondapak C_{18} (30 × 0.78 cm column, 10 μm particle size, Waters Associates) where ABA-GE eluted after PA. Daie et al. (1984) reported using a two-step gradient consisting of 0% to 9.5% ethanol in 0.1 N acetic acid in 2 min followed immediately by a second gradient to 47.5% ethanol in 0.1 N acetic acid in 25 min (flow rate 2 ml min^{-1}, R Sil C_{18} column 5 × 1 cm, 10 μm particle size, Alltech Associates, Arlington Heights, IL, USA)

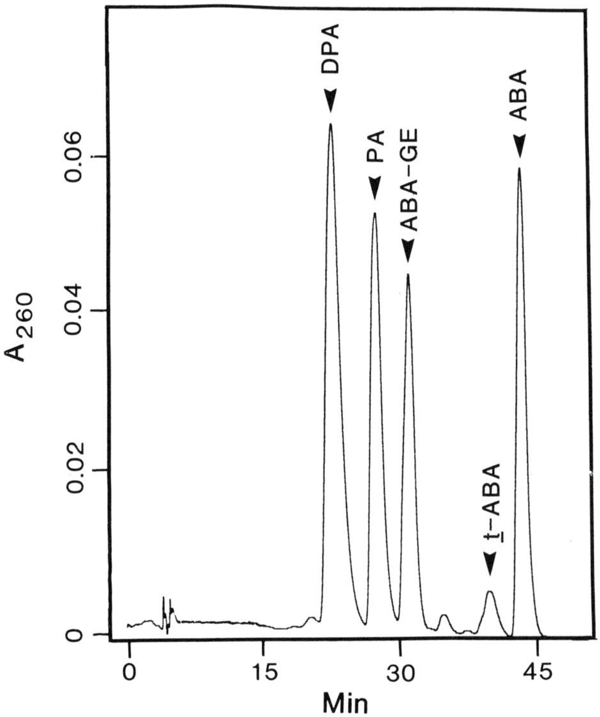

Fig. 4. Separation by high performance liquid chromatography of abscisic acid and its catabolites. Standards shown are DPA (22.7 min), PA (27.6 min), ABA-GE (31.2 min), *t*-ABA (40.0 min), and ABA (43.3 min). Analysis was performed with a μBondapak C_{18} column (Waters Associates) with a gradient of 5% to 30% ethanol in 1% acetic acid in 40 min (flow rate 2.5 ml min^{-1}). There was a 3-min delay before the beginning of the gradient to allow the injection loop to empty

which partially resolved DPA, PA, and ABA-GE. Vaughan and Milborrow (1984) separated the catabolites of ABA with ethanol:water:acetic acid (84:504:1, v:v:v) run isocratically for 22 min followed by a linear gradient to 158:508:1 in 15 min (flow rate 4 ml min^{-1}) using a μBondapak C_{18} reverse phase column (30 × 0.78 cm, Waters Associates).

If separation of ABA-GE and PA is difficult, the fraction containing both PA and ABA-GE may be collected and the two compounds can be further resolved by another HPLC step. To separate these two compounds, Setter et al. (1981) used a μBondapak-NH_2 column (10 µm particle size, Waters Associates). Samples were injected in 5 ml of solvent consisting of 0.1 N acetic acid and water: ethanol:acetonitrile (5:95:900, v:v:v) run isocratically for 10 min followed by a linear gradient to ethanol:water (95:5, v:v) in 10 min. Alternatively, the fraction containing PA and ABA-GE may be collected and the ester linkage of ABA-GE hydrolyzed with 2 M NH_4OH at 60 °C for 2 h. After drying, the sample may be methylated and analyzed by GC, or PA and the released ABA can be separated and purified by HPLC.

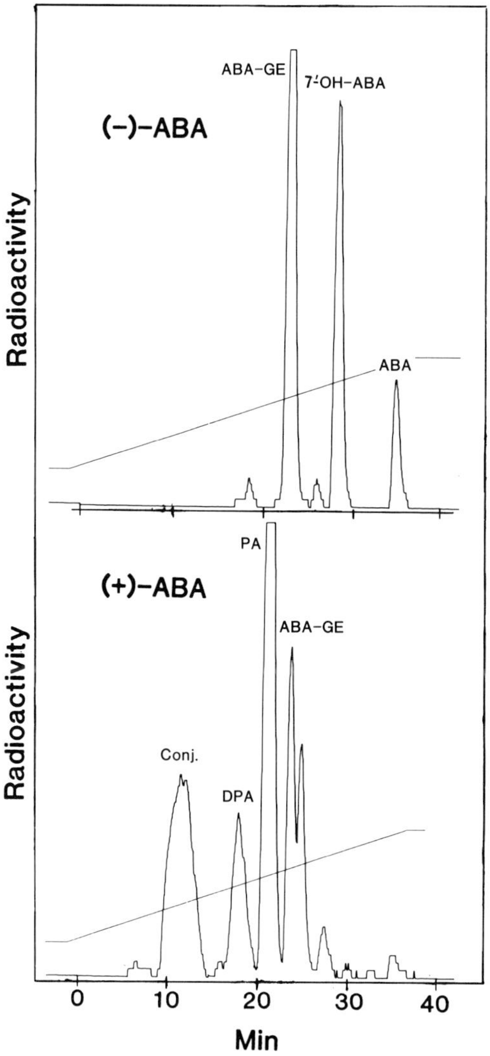

Fig. 5. Comparison of the metabolism of (+)- and (−)-[^{14}C]-abscisic acid in leaves of *Xanthium strumarium*. Leaves were fed 5.5 µg of (−)- or (+)-[2-^{14}C]-ABA (specific activity 4 mCi mmol^{-1}) for 2 days. Analysis was performed as in Fig. 4, except that a gradient of 10 to 35% ethanol and a 2.5 min delay were used. Radioactivity was detected with a Flo-One radioactive flow detector (Radiomatic Instruments and Chemical Co., Inc., Tampa, FL, USA). Full scale for radioactivity: 10^4 dpm. The two main catabolites of (−)-ABA were (−)-ABA-GE (23.6 min, 59%) and 7'-OH-(−)-ABA (28.9 min, 29%). (+)-ABA was metabolized to PA (21.5 min, 44%), DPA (18.2, 9%), conjugates of PA and DPA (12 min, 22%), ABA-GE (24.0 min, 12%), and two unknown catabolites (25.1 min, 7%; 27.7 min, 2%)

For further analysis by GC-ECD, GC-MS, RIA, or EIA, fractions are collected based on the retention times of standards. Free acid catabolites of ABA (such as PA and DPA) may be derivatized with ethereal diazomethane and analyzed by GLC-ECD without further workup. Conjugated catabolites, such as ABA-GE, must first be hydrolyzed, and then derivatized before further analysis. If more purification of the sample is needed, one may perform a further HPLC step utilizing either normal or reverse phase HPLC. For example, Me-ABA can be further purified with a µPorasil column (Waters Associates), using a 10 to 60% ethyl acetate in hexane gradient (flow rate 2 ml min^{-1}) in 10 min (RA Creelman, JAD Zeevaart, unpublished results, 1986).

In many studies radioactive ABA is used to determine the metabolism of endogenous ABA (Setter et al. 1981; Daie et al. 1984; Loveys and Milborrow 1984;

Vaughan and Milborrow 1984; Vaughan 1986). HPLC allows the easy determination and isolation of catabolites of radioactive ABA with the use of an on-line liquid scintillation counter or by manual collection and subsequent counting. These techniques are superior to analysis of TLC zones, because (a) resolution is greater, and (b) there is less handling of radioactive materials. However, because synthetic radioactive ABA is a racemic mixture, the observed catabolism may not reflect normal endogenous metabolism of ABA. An example of this technique is found in Fig. 5 showing the catabolism of (−)- and (+)-[^{14}C]-ABA. Clearly, the catabolism of (−)- and (+)-[^{14}C]-ABA is quite different. A major difference is the appearance of a peak at retention time 28.9 min when (−)[^{14}C]-ABA was fed which was identified as 7′-hydroxy-(−)-ABA (Boyer and Zeevaart 1986). Lehmann et al. (1983) reported the formation of this compound after feeding [^{14}C]-ABA (no stereochemistry reported). However, the work of Boyer and Zeevaart (1986) unequivocally showed that 7′-hydroxy-(−)-ABA is a catabolite of (−)-ABA. Thus, all studies dealing with ABA catabolism should state the stereochemistry of ABA fed and of the catabolites isolated.

Conversion of radioactive ABA (either ^{3}H or ^{14}C) to vomifoliol (Fig. 3) or other potential C_{13} catabolites will not be detected by the techniques described above. This is because label will be lost if cleavage of the side chain of ABA occurs to form vomifoliol (^{14}C is at C-2, ^{3}H in the side chain methyl group will be lost by exchange). Therefore, to determine if these compounds are catabolites of ABA, radioactive ABA must be used containing label in a position that will not be lost during the conversion to vomifoliol.

4 Concluding Remarks

Reverse phase HPLC is an excellent method to quickly and reproducibly separate ABA and its catabolites. However, HPLC is not the preferred method for quantification of ABA, since the method is neither very sensitive nor selective. Following HPLC, ABA can be measured by one of several very sensitive methods, such as GC-ECD, EIA, or RIA, as long as identity is verified by GC-MS.

With the availability of sensitive methods, it is now possible to quantitate ABA at the cellular rather than at the gross organ level. It is hoped that these refined methods will contribute to an unraveling of the role(s) of ABA in plant growth and development.

Acknowledgments. Our research reported here as well as the preparation of this chapter were supported by the National Science Foundation through grant PCM 83-14321 and by the United States Department of Energy under Contract DE-AC02-76 ER01338.

References

Addicott FT, Lyon JL (1969) Physiology of abscisic acid and related substances. Annu Rev Plant Physiol 20:139:164

Addicott FT, Lyon JL, Ohkuma K, Thiessen WE, Carns HR, Smith OE, Cornforth JW, Milborrow BV, Ryback G, Wareing PF (1968) Abscisic acid: a new name for abscisin II (dormin). Science 159:1493

Assante GL, Merlini G, Nasini G (1977) (+)-Abscisic acid, a metabolite of the fungus *Cercospora rosicola*. Experientia 33:1556

Boyer GL, Zeevaart JAD (1982a) Metabolism of abscisic acid in *Xanthium strumarium*. Plant Physiol Suppl 69:77

Boyer GL, Zeevaart JAD (1982b) Isolation and quantitation of β-D-glucopyranosyl abscisate from leaves of *Xanthium* and spinach. Plant Physiol 70:227–231

Boyer GL, Zeevaart JAD (1986) 7′-Hydroxy (−)-*R*-abscisic acid: a metabolite of feeding (−)-*R*-abscisic acid to *Xanthium strumarium*. Phytochemistry 25:1103–1105

Boyer GL, Milborrow BV, Wareing PF, Zeevaart JAD (1986) The nomenclature of abscisic acid and its metabolites. In: Bopp M (ed) Plant growth substances 1985. Springer, Berlin Heidelberg New York Tokyo, pp 99–100

Brenner ML, Burr B, Burr F (1977) Correlation of genetic vivipary in corn with abscisic acid concentration. Plant Physiol Suppl 63:36

Brenner ML (1981) Modern methods for plant growth substance analysis. Annu Rev Plant Physiol 32:511–538

Cornforth JW, Mallaby R, Ryback G (1968) Synthesis of (±)-[2-^{14}C]abscisic acid. J Chem Soc (C) 1565–1568

Cornish K, Zeevaart JAD (1984) Abscisic acid metabolism in relation to water stress and leaf age in *Xanthium strumarium*. Plant Physiol 76:1029–1035

Crouch ML, Sussex IM (1981) Development and storage protein synthesis in *Brassica napus* L. in vivo and in vitro. Planta 153:64–74

Crouch ML, Tenbarge K, Simon A, Finkelstein R, Scofield S, Solberg L (1985) Storage protein mRNA levels can be regulated by abscisic acid in *Brassica* embryos. In: van Vloten-Doting L, Groot GSP, Hall TC (eds) Molecular form and function of the plant genome. Plenum Press, New York, pp 555–566

Daie J, Wyse R, Hein M, Brenner ML (1984) Abscisic acid metabolism by source and sink tissues of sugar beet. Plant Physiol 74:810–814

Dathe W, Sembdner G (1982) Isolation of 4′-dihydroabscisic acid from immature seeds of *Vicia faba*. Phytochemistry 21:1798–1799

Davies WJ, Mansfield TA (1983) The role of abscisic acid in drought avoidance. In: Addicott FT (ed) Abscisic acid. Praeger Press, New York, pp 237–268

Dörffling K, Tietz D (1983) Methods for the detection and estimation of abscisic acid and related compounds. In: Addicott FT (ed) Abscisic acid. Praeger Press, New York, pp 23–78

Dure L, Galu G, Chlan C, Pyle J (1983) Developmentally regulated gene sets in cotton embryogenesis. In: Goldberg R (ed) Plant molecular biology. Liss, New York, pp 331–342

Gaskin P, MacMillan J (1968) Identification and estimation of abscisic acid in a crude plant extract by combined gas-chromatography-mass spectrometry. Phytochemistry 7:1699–1701

Gillard DF, Walton DC (1976) Abscisic acid metabolism by a cell-free preparation from *Echinocystis lobata* endosperm. Plant Physiol 58:790–795

Harrison MA, Walton DC (1975) Abscisic acid metabolism in water stressed bean leaves. Plant Physiol 56:250–254

Hasegawa S, Poling SM, Maier VP, Bennett RD (1984) Metabolism of abscisic acid: bacterial conversion to dehydrovomifoliol and vomifoliol dehydrogenase activity. Phytochemistry 25:2769–2771

Hirai N, Fukui H, Koshimizu K (1978) A novel abscisic acid metabolite from seeds of *Robinia pseudoacacia*. Phytochemistry 17:1625–1627

Hirai N, Koshimizu K (1983) A new conjugate of dihydrophaseic acid from avocado fruit. Agric Biol Chem 47:365–371
Ho T-HD (1983) Biochemical mode of action of abscisic acid. In: Addicott FT (ed) Abscisic acid. Praeger Press, New York, pp 147–169
Ihle JN, Dure L (1970) Hormonal regulation of translation requiring RNA synthesis. Biochem Biophys Res Comm 38:995–1001
Jacobsen JV (1983) Regulation of protein synthesis in aleurone cells by gibberellin and abscisic acid. In: Crozier A (ed) The biochemistry and physiology of gibberellins, vol 2. Praeger Press, New York, pp 159–188
Karssen CM, Laçka E (1986) A revision of the hormone balance theory of seed dormancy: studies on gibberellin and/or abscisic acid deficient mutants of *Arabidopsis thaliana*. In: Bopp M (ed) Plant growth substances 1985. Springer, Berlin Heidelberg New York, pp 315–323
Knox JP, Galfre G (1986) Use of monoclonals to separate the enantiomers of abscisic acid. Anal Biochem 155:92–94
Lehmann H, Miersch O, Schütte HR (1975) Synthese von Abscisyl-β-D-glucopyranosid. Z Chem 15:443
Lehmann H, Preiss A, Schmidt J (1983) A novel abscisic acid metabolite from cell suspension cultures of *Nigella damascena*. Phytochemistry 22:1277–1278
Lehmann H, Repke H, Gross D, Schütte HR (1973) Darstellung von Abscisinsäure-2-^{14}C. Z Chem 13:255–256
Loveys BR, Downton JS (1979) Abscisic acid in crushed grape berries during fermentation and in wine during storage. Am J Enol Viticult 30:232–235
Loveys BR, Milborrow BV (1984) Metabolism of abscisic acid. In: Crozier A, Hillman JR (eds) The biosynthesis and metabolism of plant hormones. Cambridge University Press, London, pp 71–104
Milborrow BV (1967) The identification of (+)-abscisin II [(+)-dormin] in plants and measurement of its concentration. Planta 76:93–113
Milborrow BV (1969) Identification of "metabolite C" from abscisic acid and a new structure for phaseic acid. Chem Commun 966–967
Milborrow BV (1983) The reduction of ($\pm$)-[2-^{14}C] abscisic acid to the 1′,4′-trans-diol by pea seedlings and the formation of 4′-desoxy ABA as an artifact. J Exp Bot 34:303–308
Milborrow BV, Mallaby R (1975) The occurrence of methyl (+)-abscisate as an artifact of extraction. J Exp Bot 26:741–748
Milborrow BV, Vaughan GT (1982) Characterization of dihydrophaseic 4′-*O*-β-D-glucopyranoside as a major metabolite of abscisic acid. Aust J Plant Physiol 9:361–372
Moore R, Evans ML (1986) How roots perceive and respond to gravity. Am J Bot 73:574–587
Moore R, Smith JD (1985) Graviresponsiveness and abscisic acid content of roots of carotenoid-deficient mutants of *Zea mays* L. Planta 164:126–128
Neill SJ, Horgan R, Heald JK (1983) Determination of the levels of abscisic acid-glucose ester in plants. Planta 157:371–375
Raikhel NV, Palevitz BA, Haigler CH (1986) Abscisic acid control of lectin accumulation in wheat seedlings and callus cultures. Plant Physiol 80:167–171
Raschke K (1975) Stomatal action. Annu Rev Plant Physiol 26:309–340
Roberts DL, Heckman RA, Hege BP, Bellin SA (1968) Synthesis of (*RS*)-abscisic acid. J Org Chem 33:3566–3569
Setter TL, Brun WA, Brenner ML (1981) Abscisic acid translocation and metabolism in soybeans following depodding and petiole girdling treatments. Plant Physiol 67:774–779
Smith JD, McDaniel S, Lively S (1978) Regulation of embryo growth by abscisic acid in vitro. Maize Genet Coop News Lett 52:107–108
Sondheimer E, Galson EC, Tinelli E, Walton D (1974) The metabolism of hormones during dormancy. IV. The metabolism of (S)-2-^{14}C abscisic acid in ash seed. Plant Physiol 54:803–808

Sussex I (1975) Growth and metabolism of the embryo and attached seedling of the viviparous mangrove *Rhizophora mangle*. Am J Bot 62:948–953

Takahashi S, Oritani T, Yamashita K (1986) Total synthesis of (±)-methyl phaseates. Agric Biol Chem 50:1589–1596

Vaughan GT (1986) Studies on the metabolism of abscisic acid. Thesis, University of New South Wales, Kensington

Vaughan GT, Milborrow BV (1984) The resolution by HPLC of *RS*-[2-^{14}C]Me 1′,4′-cis-diol of abscisic acid and the metabolism of (−)-*R*- and (+)-*S*-abscisic acid. J Exp Bot 150:110–120

Walton DC (1980) Biochemistry and physiology of abscisic acid. Annu Rev Plant Physiol 31:453–489

Walton D, Wellner R, Horgan R (1977) Synthesis of tritiated abscisic acid of high specific activity. Phytochemistry 16:1059–1061

Wang TL, Griggs P, Cook S (1986) Immunoassays for plant growth regulators – A help or hindrance? In: Bopp M (ed) Plant growth substances 1985. Springer, Berlin Heidelberg New York, pp 26–34

Wareing PF (1978) Abscisic acid as a natural growth regulator. Phil Trans Roy Soc Lond B 284:483–498

Wilkins MB (1978) Gravity sensing guidance mechanisms in roots and shoots. Bot Mag Spec Issue 1:255–277

Wilkins MB (1984) Gravitropism. In: Wilkins MB (ed) Advanced plant physiology. Pitman, London, pp 163–185

Williamson JD, Quatrano RS, Cuming AC (1985) E_m polypeptide and its messenger RNA levels are modulated by abscisic acid during embryogenesis in wheat. Eur J Biochem 152:501–507

Zeevaart JAD (1977) Sites of abscisic acid synthesis and metabolism in *Ricinus communis* L. Plant Physiol 57:788–791

Zeevaart JAD (1979) Chemistry and biological aspects of abscisic acid. In: Mandava NB (ed) Plant growth substances. ACS symposium series, no 111. Washington, DC, American Chemical Society, pp 99–114

Zeevaart JAD, Milborrow BV (1976) Metabolism of abscisic acid and the occurrence of *epi*-dihydrophaseic acid in *Phaseolus vulgaris*. Phytochemistry 15:493–500

The Determination of Abscisic Acid by High Performance Liquid Chromatography

J. M. ROBERTSON

1 Introduction

In the last 10 years, high performance liquid chromatography (HPLC) has largely replaced the older forms of liquid chromatography, gravity column and thin layer chromatography as the method of choice for separating plant hormones. Apologists for HPLC point out the shorter separation times, higher resolution, and greater convenience of HPLC compared to alternative methods. While the advantages of HPLC are veritable, they are not always realized. Separating abscisic acid (ABA) from a plant extract by HPLC is relatively easy, but achieving separations of high quality may prove difficult.

This rapid increase in the use of HPLC has not come without some attendant problems. Upon first considering HPLC as a tool for ABA research, the prospective user will consult the literature for the "preferred" method; seeking common observances but finding only nonconformity, he may well complain. "Everyone does something different".

"Why are many different methods used? Which method is best? How does one choose?". These questions are the genuine concerns of the ABA researcher and the points of resolution for this discussion.

1.1 HPLC and ABA

1.1.1 Why Have Many Different Methods Been Used?

The answer to that question is quite simple: many different methods have been used because many different methods work. Why then has there been no consensus? Several reasons suggest themselves: recent advances in HPLC, particularly in the area of column packings, are in part responsible, as are the preferences of individual researchers, who, having experienced success with one technique, are reluctant to try others. But a more fundamental and compelling reason for the continued use of different methods lies closer to the very nature of plant analysis: no two separation problems are exactly alike. Different problems require different solutions.

1.1.2 Which Method Is Correct? How to Choose

The correct choice of method depends upon the demands of the research: the tissue system from which ABA is to be separated, the amount of tissue available for analysis, and the importance assigned to HPLC in the overall purification scheme are all factors in the choice of method. No one method or set of operating conditions will apply to all situations; however, for each situation a method may be found. The lack of conformity, that seems a drawback at first glance, actually constitutes the greatest strength of HPLC, versatility.

1.1.3 General Approach to HPLC of ABA

The common approach to setting up an HPLC system for separating ABA is to make an attempt, and then adjust the system until it works. This can be a very time-consuming task. Unfortunately, this article will not eliminate the need for that approach, but it may help shorten the process. There are no real shortcuts to setting up a system, but there are many wrong turns which can be avoided. The successful separation of ABA by HPLC depends upon the ability of the operator to adapt the system to the task at hand.

The researcher should proceed in an orderly fashion with a clear sense of purpose. This requires an understanding of the chromatographic process and of the ABA molecule, its chemical properties and biological occurrence. Above all, the user must keep in mind what he is attempting to achieve by the use of HPLC.

HPLC is a powerful technique, but it has its limitations; and like any powerful technique, it places strict demands upon those who employ it. What follows are some general considerations of HPLC, a discussion of the problems peculiar to ABA chromatography, and a generalized scheme for separating ABA.

1.2 HPLC General

1.2.1 Theoretical Considerations

Much of the separation process is under the control of the operator. This section introduces some basic theoretical considerations which can be applied to effect better chromatography. What follows is for the purpose of this argument only, and should not be used in place of a rigorous treatise on HPLC theory. For a solid background to HPLC, the reader should consult Knox et al. (1978), Simpson (1976), or Snyder and Kirkland (1979).

1.2.1.1 Retention

The retention of any compound on HPLC depends upon its distribution between the stationary and mobile phases:

$$K = C_s/C_m, \tag{1}$$

where K is the thermodynamic equilibrium constant. C_s is the concentration of solute in the stationary phase and C_m is the concentration of solute in the mobile phase. A more useful quantity is the column capacity ratio, k′:

$$k' = n_s/n_m = K(V_s/V_m), \tag{2}$$

where n_s is the number of moles of the solute in the stationary phase, and n_m is the number of moles of the solute in the mobile phase; V_s and V_m refer to the respective column volumes of stationary and mobile phases. The column capacity ratio may be determined empirically:

$$k' = (t_R - t_0)/t_0 = (V_R - V_0)/V_0, \tag{3}$$

where t_R is the retention time of the solute, t_0 is the time required to elute an unretained band, and V_R and V_0 (the column dead volume) refer to the volumes of solvent required to elute the solute and an unretained band, respectively. Two solutes must have different values of k' for any system on which they are to be separated, but this is not enough to ensure their separation.

1.2.1.2 Band Spreading

Solute bands tend to spread (become more dilute) as they migrate down an HPLC column. Column efficiency, commonly expressed as the theoretical plate number, N, is a relative measure of resistance to band spreading:

$$N = 16(t_R/t_w)^2. \tag{4}$$

Under fixed operating conditions, the quantity, N, is virtually constant for different bands in a chromatogram; hence, Eq. (4) predicts that band width will increase proportionately with t_R (Snyder and Kirkland 1979, Chap. 2). More succinctly, for a given value of N, the width of a band is a function of its k' value, since $t_R = k't_0 + t_0$. The height equivalent to a theoretical plate, or simply plate height, H, measures column efficiency per unit length:

$$H = L/N = (L/16)\,(t_w/t_R)^2, \tag{5}$$

where t_w is the band width in time units. The higher the N value, the better the separation that can be expected from the column.

1.2.1.3 Resolution

Resolution, R_s, is a measure of the relative separation of two adjacent bands; it may be defined as the distance between the band centers divided by the average band width, t_w (Snyder and Kirkland 1979, Chap. 2):

$$R_s = (t_{R1} - t_{R2})/(1/2)(t_{w1} + t_{w2}). \tag{6}$$

Equation (7) relates separation to the thermodynamic and kinetic parameters of HPLC:

$$R_s = (1/4)(\alpha - 1)\,[\bar{k}'/1 + \bar{k}']N^{1/2}, \tag{7}$$

where: α (the separation factor) $= k'_2/k'_1$; $\bar{k}' = 1/2\,(k'_1 + k'_2)$; and N is the number of theoretical plates. Equation (7) is a fundamental relationship which allows the user to control separation by manipulating α, N, or k'; the three terms are roughly independent, so that the user may optimize first one then another of the terms (Snyder and Kirkland 1979, Chap. 2).

1.2.1.4 Controlling R_S by N, k′, and α

Suppose a separation has been attempt, and it is found that R_S for ABA and some interfering compound is not satisfactory. An increase in N will sharpen the bands (reduce band width); the result will be less band overlap. An increase in k′ will spread the bands from one another, but at the cost of increased t_R. An increase in α can separate the band centres, while still allowing for low k′ and high N.

Generally, smaller particles yield higher N values (Knox et al. 1978; Snyder and Kirkland 1979). For most modern HPLC applications, columns of 5–10 μm particles are used, and N values are potentially high. This means that if N is to be further increased, then it must be done by increasing L or decreasing u, the solvent velocity. L may be increased by adding another length of column with low dead volume connectors (column coupling). This will require adjustments in t_R and/or pressure. The other available alternative is to decrease u, the mobile phase flow rate. Decreasing u is a much easier and quicker correction, but increasing L yields better results and faster separation times. An increase in L is the preferred method if a large number of sample runs are to be made. Knox et al. (1978, Chap. 2) provide formulae which can be used to predict the results of changes in operating conditions.

For most ABA separations, the potentially high N values of modern HPLC packings will provide for satisfactory column efficiency. More often a change in solvent strength or solvent selectivity will correct the problem.

A change to a stronger solvent will effect the earlier elution of all bands (lower k′ values). There are advantages to keeping k′ values low: separation times are shorter, but, more importantly, early-eluting bands are narrower. Since most detectors respond to solute concentration, sensitivity decreases as k′ increases [Eq. (4)]. However, bands do separate from one other as k′ values increase. A change in solvent strength brings about a change in the equilibrium distribution constant, but not necessarily a change in the fundamental processes of separation. For example, the mobile phase is usually a mixture of two or more solvents; a change from 50% aqueous methanol to 40% aqueous methanol mobile phase on a reverse-phase C_{18} column would reduce solvent strength and increase k′ values for the various solutes, but would not change their elution order.

Separation selectivity measured as α, the separation factor, refers to the basis on which molecules of different solutes separate from one another. A change in α represents a constitutional change in the intermolecular forces which drive the chromatographic process. Replacing one or more of the components in the solvent mixture will change α. For instance, the replacement of methanol by acetonitrile as the strong solvent for a reverse-phase C_{18} column will alter the elution profile. The other, more radical, means of changing α is changing the chromatographic method, e.g., switching from reverse to normal phase, or switching from adsorption to partition chromatography.

1.2.2 Preparative Versus Analytical Operation

HPLC has been widely used both as a preparative and an analytical tool in ABA research. Preparative and analytical techniques differ from one another primarily in purpose. In preparative HPLC, the object is to purify large amounts of sample for subsequent analysis. Large bore columns and large particle sizes best suit this purpose. The sample capacity of the column is often deliberately exceeded. In analytical HPLC, the purpose is to identify tentatively a compound by its behavior on the column. Determination may be qualitative or quantitative; however, the sample capacity of the column must not be exceeded. The first effect of exceeding the sample capacity is an increased band broadening, thereby lowering resolution and rendering quantitative determinations suspect. If the overloading is too great, t_R values may change as well. It was once thought that changes in k′ were a good indication of sample capacity limits. It is apparent, however, that band broadening may begin at much lower sample loads than do changes in k′ (Knox et al. 1978, Chap. 3). Analytical HPLC is generally done on small bore columns (ca. 0.4 cm i.d.) packed with small porous microparticles.

With plant hormone analysis, the distinctions between preparative and analytical break down. Most often the researcher will wish to purify the sample for further analysis, while at the same time making a first tentative identification of ABA. Semi-preparative columns of small particles and intermediate bore (0.6–1.0 cm i.d.) provide for reasonably high sample capacity, 10–14 mg dry wt. These columns, while providing for an adequate sample, also allow for relatively short separation times and workable peak volumes for identification. Peak volume, the volume of solvent in which the solute band elutes, increases with column diameter (Knox et al. 1978, Chap. 8). Brenner (1981) suggests the use of spherical 7 or 10 mm particles in 1.0 cm × 25 m C_{18} reverse-phase semi-preparative columns for the initial clean-up of plant extracts. He reports that these columns will accommodate crude extract from 5–10 g fr.wt. tissue.

For ABA separations, a first HPLC step does not usually meet all the requirements for an analytical separation. The column is usually overloaded; however, it should not be overloaded to the point where K′ values begin to decline. In this way, a valuable piece of information is preserved.

1.2.3 Elution

Isocratic elution will accomplish most ABA separations. There are, however, situations where gradiant elution may improve the separation. Using gradient elution is, in fact, an alternative to optimizing k′. As solvent strength increases in isocratic elution, resolution decreases, while sensitivity increases. The final choice of solvent strength, then, is a compromise between resolution and sensitivity (Snyder 1980).

The use of gradient elution for ABA separation may serve one of four purposes:

1. to separate several compounds of interest over a wide range of k′ values;
2. to improve the resolution of ABA from early-eluting compounds;

3. to speed up a separation where R_S for ABA is great but k′ large;
4. to clean up a column rapidly, following the isocratic elution of ABA.

The first use is of importance primarily for the separation of ABA from its metabolites or from other compounds of interest such as the other plant hormones. In isocratic elution, bands tend to broaden as t_R increases, so that for a mixture of compounds weakly retained bands elute in sharp peaks, but strongly retained bands elute in broad dilute peaks. Moreover, these late-eluting bands often suffer from excessive band tailing. A linear solvent strength (LSS) gradient will produce bands of virtually the same width, t_w, while reducing separation time and correcting band tailing. The choice of the correct gradient is made from an assessment of the extract in much the same way as is done for isocratic elution. Changes in t_R (k′) are accomplished by increasing the steepness of the gradient, i.e., by increasing the rate of increase of the strong solvent. As in isocratic elution, a change in solvent strength does not greatly change solvent selectivity or elution order. Although not usually necessary for ABA separations, gradients may have different shapes as well as different steepnesses. The guidelines for establishing the correct shape and steepness of a gradient are too detailed for this presentation; the reader is referred to Snyder and Kirkland (1979, Chap. 16) and Snyder (1980).

The second situation where gradient elution may be of value is where ABA elutes in a sharp peak (high N and low k′), but is poorly resolved from other early-eluting bands. Decreasing solvent strength in an isocratic elution will increase t_R and decrease sensitivity. For example, elution with 50% MeOH on a C_{18} reverse phase column may yield a desirable k′ value (≈ 5), but poor resolution for ABA. A gradient elution from about 20% to 100% MeOH over 30 min may solve this problem more satisfactorily than would a change in solvent strength in an isocratic elution. The weakly retained compounds would elute in the early part of the gradient, well before ABA. Only experiment will determine whether a gradient will be of more value than a change in isocratic solvent strength.

The third situation is one which is, in a sense, the obverse of situation 2 above. An isocratic elution provides high R_S of ABA, but at an awkwardly long t_R. An increase in the flow rate will reduce t_R but at the cost of N. If the reduction in N is not too severe, this method is the simplest. However, if this approach is not workable, a gradient may serve to lower t_R while maintaining adequate R_S.

The fourth situation is very simple. An isocratic elution may provide for satisfactory separation of ABA, but there remain a number of late-eluting compounds which require long separation times. A steep gradient or a step gradient to a stronger solvent, following the elution of ABA, will reduce t_R of these compounds. The column is then ready for the next injection much sooner. This use of gradient elution is of most value with BPC packings where column equilibration is rapid (Jones et al. 1980). If the time required to equilibrate the column to isocratic conditions is excessive, this step may be unpractical.

There are some problems with gradient elution, and it should not be used without a clear need to do so. Varying solvent strength further complicates an already complex situation: retention is difficult to predict; moreover, achieving consistency in the gradient may be difficult in all but the best HPLC equipment.

Gradient elution should be avoided for quantitative trace analysis (Snyder and Kirkland 1979, Chap. 13; Lawrence 1981, p. 130). Where precise quantitative data are required, baseline drift and poor reproducibility are two major drawbacks of gradient elution.

1.2.4 The Different Modes of HPLC

The modes of chromatography differ from one another in the thermodynamic forces which control the retention of solutes. With a change from one mode to another, then, one would expect a great change in separation selectivity. Several modes of HPLC produce satisfactory separation of ABA. This situation presents the ABA researcher with a means for radically changing α for difficult separations. The different methods are often complementary: one method which removes certain types of compounds can be followed by a second method which selectively removes other compounds.

1.2.4.1 Partition or Liquid-Liquid Chromatography (LLC)

In LLC, solute molecules are partioned between two immiscible phases: one the mobile phase, and the other, the stationary phase coated onto a support. Separation is on the basis of the type and number of substituent groups and by molecular weight differences (Snyder and Kirkland 1979). LLC requires some awkward preparations which have severely reduced its use, not only for ABA research, but for most other HPLC applications as well. The mobile phase must be presaturated with stationary phase and passed through a precolumn to prevent column stripping. Changes in the mobile phase are time-consuming and gradient elution difficult. Plant hormone researchers have virtually ignored LLC, largely because their first real interest in HPLC coincided with the advent of bonded-phase packings, which has relegated LLC to the status of a last resort. Nevertheless, the wide array of stationary phases available with LLC offers a potentially broad range of selectivity; LLC may hold as yet unexplored value for ABA research.

Crozier and Reeve (1977) used a 10 × 450 mm Partisil 10 column coated with 40%, 0.5 M formic acid and eluted it with 55% ethyl acetate in hexane as a preparative HPLC step. They reported very high sample capacity with this system (in excess of 100 mg of a multi-component mixture), which makes it a valuable tool for purifying large samples or relatively crude extracts.

1.2.4.2 Adsorption or Liquid-Solid Chromatography (LSC)

The basis of separation of LSC is the selective adsorption of the components of a mixture onto active sites on the surface of an adsorbent, usually silica gel. LSC separates compounds of different chemical type and compounds which differ in the number of functional groups (Snyder and Kirkland 1979). LSC is versatile in that it allows for a wide range of solvent systems. However, for results of high quality, the column must be conditioned with polar modifiers (Knox et al. 1978,

p. 24). This can be a somewhat bothersome task; and for that reason, bonded phase packings have largely replaced adsorbents in HPLC.

Snyder and Kirkland (1979) list LSC as the preferred method for preparative HPLC. Curiously, LSC has been ignored for preparative ABA separations but has been used frequently as an analytical method.

Chia et al. (1977) found that a 10-μm μ-Porasil (Waters Assoc.) column eluted with 15% acetonitrile in chloroform acidified with 0.17 N acetic acid gave sharp narrow peaks at a low t_R. They reported a lower detection limit of 500 pg ABA with a linear capacity to 20 μg (very low). Cargile et al. (1979) used the same column type with 7% acetonitrile in ethylene chloride mobile phase for the separation of ABA methyl ester, detection limit was 20 ng. Durley et al. (1982) used an Ultrasphere Si column, 5 μm particles (Beckman) eluted with chloroform : acetonitrile : acetic acid (94:5:1), detection limit 1 ng. Fiscus et al. (1982) used a μ-Porasil column eluted with a 33.3% acetonitrile/chloroform mixture presaturated with 0.17 N acetic acid.

1.2.4.3 Size Exclusion Chromatography (SEC)

In a well-designed SEC system, retention is on the basis of one single parameter, molecular size. This is of particular advantage for the determination of ABA, since the separation limits the possible number of compounds in a peak and provides some information about their structure (see Reeve and Crozier 1980).

Crozier et al. (1980) report on the performance of a 30 × 0.8 cm μspherogel column (Altex Scientific) with a nominal pore size of 50 Å eluted with tetrohydrofuran. Because of the high sample capacity and the short elution times with SEC, they recommend this technique for the purification of relatively crude extracts. Peak capacity on SEC is low; as a result, the degree of purification on SEC does not compare favorably with that of other methods, e.g., RPLC, LSC, LLC. The appeal of SEC lies in the information it provides (see Sect. 1.3.3).

1.2.4.4 Ion Exchange Chromatography (IEC)

Ion exchange was the first HPLC method widely applied to ABA separations (Carnes et al. 1974; Düring 1977; Düring and Bachman 1975; Quebedeaux et al. 1976; Sweetser and Vatvars 1976). Early IEC packings were unstable and of low efficiency; they showed poor reproducibility and had low sample capacity. Sample retention on IEC is a combination of partition and ion exchange effects. "Solvent strength and selectivity depend on several variables: the type and concentrations of buffer ions and other salts, the pH, and the type and concentration of added organic solvents" (Snyder and Kirkland 1979, Chap. 10). All this makes for separations which are difficult to predict and control. Ion exchange soon lost favor as other methods emerged.

Modern packings of high ion exchange capacity and high efficiency have eliminated some of the drawbacks of IEC, but the technique is virtually out of use for ABA separations. Nevertheless, IEC is a very flexible method, which allows for a great deal of manipulation of the stationary and mobile phases; IEC can be kept in mind should other methods fail.

Both anion exchange (Düring and Bachman 1975) and cation exchange (Sweetser and Vatvars 1976; Quebedeaux et al. 1976) have been used for the separation of ABA.

1.2.4.5 Bonded-Phase Chromatography (BPC)

Bonded-phase packings refer to stationary phases which are covalently bound to a silica gel support. There remains some debate about the retention mechanism on BPC. However, as pointed out by Snyder and Kirkland (1979, Chap. 7), it is convenient to consider it as an LLC process: "This leads to reasonable predictions of retention as a function of solute structure and mobile phase composition" (Snyder and Kirkland 1979, Chap. 7). Separation is on the basis of type and number of functional groups within the solute molecule (Snyder and Kirkland 1979, Chap. 7). BPC offers some attractive advantages over other methods. Several BPC stationary phases of differing polarities provide for a wide choice of separation selectivity. Reverse and normal phase separations are possible. Setting up a separation on a BPC column is usually straightforward, demanding little special knowledge on the part of the operator. Because they are stable and do not require a precolumn, BPC packings readily accommodate gradient elution. Moreover, BPC columns equilibrate rapidly to a change in conditions (Engelhardt 1979, pp. 127, 148), thus column regeneration time is short (Jones et al. 1980).

The vast majority of HPLC operations are done on BPC columns, with one method, reverse-phase BPC (RPLC), accounting for an estimated 80% of all HPLC separations (Krstulovic and Brown 1982). Octadecylsilane (C_{18}) is the prevalent choice of stationary phase for ABA separations (Arteca et al. 1980; Cargile et al. 1979; Chia et al. 1977; Creelman and Zeevaart 1984; Durley et al. 1978, 1982; Hubick et al. 1986; Gaff and Loveys 1984; Moore and Smith 1984; Setter et al. 1981; Taylor et al. 1981; Vaughan and Milborrow 1984; Zeevaart 1980). Octylsilane (C_8) has also been used (Jensen et al. 1986). Glacial acetic acid ($\approx 1\%$) in water is the usual choice for the weak solvent with ethanol, methanol, or acetonitrile as the strong solvent.

Normal phase BPC, though not as common, complements RPLC. Chia et al. (1978), Jensen et al. (1986), and Zeevaart (1980) used normal phase NH_2 BPC as an analytical step, following purification by RPLC.

1.2.5 Detection

Of the several HPLC detectors available, only ultraviolet absorption (usually at 254 nm) has found widespread use for ABA studies. Other methods either lack sensitivity or are not applicable to the detection of ABA. Although sensitive to ca. 1 ng in optimized systems, UV absorption is not highly selective. Many compounds absorb in the UV range, hence, UV absorption reveals little about the compounds it detects, and consequently little about the purity of a putative ABA peak. With most HPLC methods then, the only information provided is retention time. This is a severe limitation of HPLC, as will be explained in Sect. 1.3.3.

Anderson (1985) used dansyl hydrazine to form a fluorescent hydrazone derivative of ABA which detected on a fluorometer. This technique holds considerable promise, since fluorescence detection is more selective and potentially more sensitive than UV detection.

The next major improvement in plant hormone analysis will be the development of a practicable on-line mass spectrometry detector for HPLC. Mass spectrometry is highly selective and yields more information than any other analytical technique. Unfortunately HPLC-MS still lacks sensitivity. Moreover, even should the sensitivity problems be solved, the cost of HPLC-MS will be prohibitive for many labs. On-line HPLC-MS is progressing (Games 1983); the advant of microbore columns for this purpose is promising (Ogan 1984).

On-line radio activity (R.A.) detectors are useful for monitoring samples spiked with [^{3}H] or [^{14}C] ABA; they are especially valuable for picking up ABA metabolites from the extracts of plants pre-treated with labeled ABA. However, since it is easy to collect HPLC fractions and take aliquots for liquid scintillation counting, the chief appeal of R.A. detectors is their convenience. If an R.A. detector is used, care should be taken to avoid adding to extracolumn band broadening effects (see Snyder and Kirkland 1979; Knox et al. 1978).

1.3 ABA General

1.3.1 The ABA Molecule

Abscisic acid is weak acid (pk = 4.8); its solubility and consequently its retention on many HPLC systems vary with pH. The dissociated and undissociated forms of ABA will, if run together on the same column, establish an undesirable secondary equilibria effect that results in band tailing or peak spliting. Generally, ion suppression techniques prevent this problem. Buffers or, more commonly, glacial acetic acid are used to reduce pH to $\leqq 3.0$.

The geometric isomers are interconvertible in light. As well as contributing to losses, interconversion will affect resolution. Although separable on HPLC, the two isomers usually run close together, and the trans-trans isomer often appears as a leading or trailing shoulder on the ABA peak. All extraction and clean-up procedures should be done in dim light. Covering the fraction collector is also good practice if further analysis is to follow.

Treatment with cold ethereal diazomethane derivatizes ABA to its methyl ester (ABA-ME). This is the usual practice for detection by gas chromatography-electron capture (GC-ECD) or gas chromatography-mass spectrometry (GC-MS). Derivitization is not necessary for HPLC, but it does provide the operator with another means of manipulating the basis of separation. Several researchers (Cargile et al. 1979; Gaff and Loveys 1984; Rodriguez and Sanchez Tames 1985; Vaughan and Milborrow 1984) chromatographed the methyl ester. This is most attractive if HPLC is to be followed by GC-ECD or GC-MS.

1.3.2 Extraction and Preparation

One attractive advantage of HPLC is that plant extracts often require very little pre-treatment before injection onto the column. Nevertheless, the wise use of some sample pre-treatment techniques may complement the HPLC process. The clean-up procedure will depend on the HPLC method, the size of the column, and the nature of the sample.

In some cases, solvent partioning may be considerably easier and faster than manipulating the HPLC system. Partitioning an aqueous extract at low pH (2.5–3.0) against an organic phase such as diethyl ether, N-butanol, or ethyl acetate (see Chia et al. 1977; Yokota et al. 1980) will extract ABA into the organic phase, thereby removing it from many more polar compounds. Partitioning the organic phase against water at high pH ($\approx$8.0) will extract ABA (now in the dissociated form) into the aqueous phase, thereby removing it from less polar compounds. Solvent partitioning may greatly improve resolution and shorten run times. Moreover, the removal of the bulk of the interfering compounds may allow for the use of columns of smaller bore, and may facilitate the elution of ABA at a lower value of k'. Both conditions reduce peak volume and, hence, increase detector response.

Sep-pak columns (either commercially packaged or prepared from bulk materials) provide a rapid means of cleaning up a crude extract; they may be used simply as traps for nonpolar compounds (Koshioka et al. 1983; Robertson et al. 1985) or as a rough chromatography step (Hubick and Reid 1982; Vaughan and Milborrow 1984; Viscier and Lewis 1983).

1.3.3 Determination of ABA

The identification of a compound by HPLC is done by comparing the migration rate of an unknown sample under certain thermodynamic conditions with the migration of a known standard under the same conditions. It is then a comparative test to determine if an unknown compound possesses some of same the qualities as the known standard (e.g., polarity or molecular weight as expressed by retention time on an HPLC column). In a closed system where it is known that only one compound can express a certain quality, the argument can be reduced to syllogistic proof, for example:

ABA and only ABA exhibits quality X,
compound A exhibits quality X,
therefore, compound A is ABA.

However, for a mixture of unknown composition, such as a plant extract, premise (1) above is not necessarily true. Furthermore, there is no way of determining whether it is true or not, for to state that no other compound exhibits quality X would require foreknowledge of the system. This problem applies to all analytical methods, not to HPLC alone.

One approach to the problem would be to expand premise (1) to include several other qualities of ABA, e.g., only ABA exhibits X.Y. and Z. and then test for each of these qualities. Clearly the same objection still applies. To achieve certainty by this method, one would have to find all the qualities of ABA in compound A, in which case the argument is tautologous; a compound which possesses all the qualities of ABA is ABA, by definition. Even if all the qualities of ABA could be tested for, this approach is unpractical.

In effect, one sets up the problem the other way around to eliminate other compounds that are not ABA. Ultimately, we advance the proposition that premise (1) is true on the probability that no other compound would test positively for qualities X.Y.Z, etc. The problem is that the probability cannot be measured, since information on the system is still lacking.

In an attempt to resolve this problem, Reeve and Crozier (1980) invoke the use of information theory to estimate the probability that premise (1) is true. They calculate the number of pieces of information required to reach a certain level of probability, based on an estimate of the total number of possible compounds in an extract.

Scott (1982, 1983) objects to this application of information theory. The arguments are too long for this discussion (see Scott 1982, 1983; Reeve and Crozier 1983), but if Scott's objections are correct, no amount of information short of all possible information will confirm premise (1). Scott (1982) argues for the superiority of spectrometric methods because they yield information on the structure of a compound. He further states that "a compound may be considered to be identified when the analytical data can be interpreted in terms of only one structure", but then adds "an alternative structural interpretation may always evade the chemist, and so it is not truly correct to speak of conclusive identifications". A carefully worked out structural analysis is very convincing, and leaves few doubters, but it is subject to the same objections as HPLC; without all possible information, there can be no proof of identification.

Intuitively, it would seem that two independent tests for ABA are better than one, three better than two, and so on, but there is no basis in logic for such a statement, and without some reliable estimate of probability, the cogency of any inductive step is unknown.

Whether or not the methods of Reeve and Crozier are accepted, the question remains: is HPLC sufficient for the identification of ABA? Despite the objections of the preceding paragraph, the writer favors further analysis by a more specific method of detection such as GC-ECD or GC-MS. A recent study (Jensen et al. 1986) found that even after a thorough two-step HPLC procedure, substantial impurities remained. This does not mean that HPLC has no use as an analytical tool; on the contrary, the authors (Jensen et al. 1986) are careful to point out that HPLC data contributed to the identification. Rather than relegating HPLC to the lower status of a purification technique, researchers should recognize its value as a part of the process of examination and corroboration of evidence, by which all identifications must be made.

1.4 Generalized Scheme for the Separation of ABA

Section 1.2 introduces the means of manipulating an HPLC system to meet the peculiar requirements of a particular separation. What is wanted is some starting point from which these modifications may proceed. The writer proposes a generalized scheme that will serve this need.

In light of the questions raised by Reeve and Crozier (1980, 1983) and Scott (1982, 1983), it is impossible to set a standard for the analysis of ABA by HPLC. A common practice is to employ a two-step HPLC separation. Separation selectivity is changed for the second step. Two objectives are met: the further purification of ABA and the corroboration (or rejection) of the first tentative identification. The rationale here is the same as that for two-dimensional gel electrophoresis or for two-dimensional thin layer chromatography. While such a strategy does not ensure purity, it does seem a reasonable compromise between exhaustive determination by every known means and an unconfirmed estimate from a single test.

The first step is semi-preparative; that is, the operating conditions are not as stringent as in the analytical mode, but are such that the operator may gather information toward the final identification of ABA. The column may be overloaded, but not to the point that k′ values begin to change. Preferably, resolution should be great enough so that ABA elutes in a single peak that can be recovered in isolation from all other peaks (this condition may prove difficult to meet). The extract then has been purified to the point that only ABA and compounds that co-chromatograph with ABA are taken for the second (analytical) step. From this semi-preparative step, the operator makes the first tentative identification of ABA (by the t_R), while at the same time greatly purifying the extract.

The second step is an analytical separation and the conditions thereof must be met: the linear capacity of the column cannot be exceeded; resolution must be great enough to determine the peak height accurately. A column of small diameter (ca. 4 mm) with a microparticle (5–10 μm) packing is used. Isocratic elution which provides for a stable baseline is preferable. If the putative ABA peak is to be further analyzed, then again, R_S should be great enough to remove the peak in isolation from interfering peaks.

1.4.1 Choosing the Columns

For the first step, the writer suggests reverse phase chromatography (RPLC) on a C_{18} bonded phase packing of 5–10 μm diameter microparticles. Preparative columns of wide column bore and large particle size are generally not necessary, and in fact a poor choice unless very large sample sizes (> 10 g fr.wt. are needed. Semi-preparative RPLC columns of intermediate bore (6–10 mm i.d.) will accommodate relatively crude extracts from 5–10 g (fr.wt), of plant tissue (Brenner 1981). If sample sizes are small (< 1 g fr.wt.), a preparative column (ca. 4 mm i.d.) may be used. Generally, the column should be no larger than is demanded by the weight of the sample, since peak volume increases as a function of column bore (Knox et al. 1978, Chap. 7). Reducing sample weight prior to injection will facilitate the use of a smaller column.

1.4.2 Typing the System

After choosing the column or columns, the system must be typed. An amount of authentic ABA (ca. 50 ng) is injected onto the C_{18} column in 300 μl of mobile phase and eluted isocratically at the recommended flow rate of the column. Methanol : water : acetic acid (50:49:1 v/v) is a good initial solvent. The acetic acid is added for ion suppression of the ABA molecule. Solvent strength is adjusted by varying the percentage of strong solvent (MeOH) to bring k′ for ABA into the range of 2–10. The system is then tested in the presence of the extract to be purified. The addition of a radioactive spike. [^{14}C] or [^{3}H] ABA, will aid in this step. If the resolution of the radioactive peak is satisfactory ($R_S \geqq 1.5$ is usually adequate), then the system is typed, and no further adjustments are required: if not, then adjustments to k′, α, and/or N are made until a satisfactory separation is obtained. Where appropriate, the easier modifications should be tried first. The following list outlines the approach to optimizing the system in the approximate order of increasing difficulty:

1. decrease flow rate – increases N and t_R;
2. decrease solvent strength – increases k′;
3. change to gradient elution – changes k′;
4. change mobile phase – changes α;
5. increase pre-treatment – increases R_S;
6. increase L – increase N;
7. derivatize ABA – changes k′ and α;
8. change the stationary phase – changes α.

An examination of the test chromatogram will suggest the appropriate modification. If the peak is nearly resolved, then a reduction in flow rate or a simple change in solvent strength may suffice. Conversely, if R_S is very high, then flow rate may be increased to shorten the separation time. Should the resolution be very poor (<0.7), then adjustments to N by an increase in L or to α by changing to another solvent system (e.g., acetonitrile in place of methanol) may be necessary. If the problem is primarily with early-eluting compounds, a linear gradient should be tried. Should all these adjustments fail to effect a satisfactory result, then a more rigorous clean-up procedure should be considered.

1.4.3 Extraction and Preparation for RPLC

Generally, the sample capacity for plant extracts on an RPLC column may be improved by removing nonpolar compounds that have low solubility in the mobile phase. These quickly tend to overload the stationary phase, thereby reducing the sample capacity. A C_{18} sep-pak following extraction in methanol : water : acetic acid (80:19:1 v/v) is suggested for this procedure. The plant material is extracted and filtered. The filtrate is then passed through the sep-pak column, and the column washed with a volume of the extraction solvent (ca. 30% of the filtrate volume). Crude extracts in methanol should be kept acidic to prevent transestrification of ABA glucosyl ester to ABA-ME (Milborrow and Mallaby 1975).

If it is found that early eluting compounds interfere with the ABA peak, it may be better to attempt to remove some of these polar compounds, in which case the extract is reduced to the aqueous phase in vacuo, and partitioned three times against diethyl ether at pH 2.5–3.0. The aqueous phase is discarded, the organic phase taken to dryness and prepared for injection onto HPLC. If it is found that the sample still retains a high concentration of nonpolar compounds, then the pH of the aqueous phase can be changed to pH 8.0 and partitioned three times against diethyl ether. The organic phase is then discarded, and the aqueous phase taken to dryness.

Phenolic compounds can be troublesome in ABA extractions: levels of phenolics are often high in severely stressed tissue; they tend to partition with ABA and may interfere on an HPLC column. If phenolics are a problem, treatment with insoluble polyvinylpyrrolidone (PVP) may be necessary. This may be done on a short column or by slurrying the extract with PVP (Saunders 1978).

Pre-treatment will depend on the composition of the extract and on the preferences of the user. The time required for prepurification must be balanced against the benefits it provides for HPLC. The writer routinely uses a C_{18} sep-pak to increase sample capacity and partitioning with diethyl ether at low pH to remove early eluting compounds that may force an increase in k′ for ABA.

Following extraction and whatever pre-treatment is necessary, the sample is taken to dryness, resuspended in the mobile phase solvent, and filtered through a 0.5 μm filter. The extract is now ready for injection.

It is difficult to make a recommendation about injection volume; injection volumes from 25–500 μl for analytical columns are common. There are several considerations here: (1) when the whole sample is to be injected, larger volumes are easier to work with and provide for better washing of vial walls; (2) sample should not be too concentrated nor too diluted; both conditions contribute to band broadening. Snyder and Kirkland (1979, Chap. 5) suggest that the injection volume should be as large as possible, but no larger than one-fourth of the peak volume. The writer prefers to resuspend the dried extract by three washes of 300 μl of mobile phase, and then filter these individually. The filtrates are combined and taken to dryness in a conical vial under an N_2 stream. The injection volume is then taken up into the injection syringe as three 100 μl washes of the vial. This step, although time-consuming, leaves less of the sample on glass walls.

Typing the system is often the most time-consuming part of the HPLC process. However, time spent in optimizing the system will be repaid by separations of high quality and short separation times. If only a few samples are to be run, it may not be necessary to fully optimize the system. For instance, should it be found that satisfactory results are obtained at high values of k′, then the time saved by faster separations may not compensate for the effort required to optimize the system at lower values of k′.

Often the objective of an experiment is to compare ABA levels in samples from different species, tissues, or treatments. These samples are also likely to differ in the number and composition of their interfering compounds. When this is the case, the system should be tested with each type of sample. One set of conditions which accommodates all samples should then be used throughout the experiment.

Following the selection of a suitable set of operating conditions, the operator must standardize the detector response. This is done by setting up a standard curve of ABA concentrations vs. peak height response [note: peak height is preferred to peak area, since it is generally less sensitive to slight changes in band broadening and baseline drift, Snyder and Kirkland (1979)]. Minimum detector response and column linear capacity may also be determined during this step.

1.4.4 The Second Column

The second HPLC step, as mentioned above, should represent a change in separation selectivity. There are several ways to accomplish such a change; three of the easiest methods are discussed here: (1) a change in the stationary phase to a normal bonded-phase NH_2 column; (2) a change in solvent selectivity; (3) a change in the ABA molecule to its methyl ester. Each provides for a significant change in the thermodynamic parameters of separation, so as to provide corroboration of the first step.

In practice, a greater array of solvent systems are used with normal-phase separations. Chia et al. (1977) used 45% acetonitrile in chloroform acidified with 0.17 N acetic acid. Zeevaart (1980) used 40% ethyl acetate containing 1% acetic acid in hexane as the initial solvent. Either of these solvent systems would serve as a good starting point. The column is then typed and the detector response set as for the first column.

The active fractions from the first column are collected, reduced to dryness in vacuo, and resuspended in the mobile phase for injection onto the second column. Quantitative analysis of ABA may now be done from the detector response, and losses calculated from liquid scintillation counting of aliquots from the active fractions.

Chia et al. (1977) suggested either changing the solvent or derivatizing ABA as alternatives to changing the stationary phase. Moore and Smith (1984) used the same C_{18} analytical column for both the first and second HPLC steps. In the second step, they replaced methanol with acetonitrile as the strong solvent, and thereby, achieved a change in α sufficient to purify ABA further. Robertson et al. (1985) adapted the method of Taylor et al. (1981) so that the same column and solvent system could be used for both steps. They eluted ABA isocratically in 32.5% acidified aqueous methanol on a C_{18} μbondapak (Waters Assoc.) analytical column (4.0 mm × 30 mm) followed by a steep linear gradient to 100% methanol. The active fraction was collected, taken to dryness, resuspended in five drops of absolute methanol, and treated with cold ethereal diazomethane. The treated extract was then run on the C_{18} column under the same conditions. ABA eluted as its methyl ester in the absolute methanol following the gradient. This method changed both the solute and the strength of the solvent in which it eluted. The final purified extract then contained only ABA and compounds which: co-chromatographed with ABA under isocratic elution, were derivatized by diazomethane, and whose derivatives co-chromatographed with ABA-ME.

The two alternatives listed above may save considerable time in switching columns. They offer a real advantage only if the sample is small enough to allow for the use of an analytical column in the first step. The first method (Moore and Smith 1984) is suited to situations where HPLC is the only means of determination. The method of Robertson et al. (1985) is of value where HPLC is to be followed by another method of determination which requires the derivatization of

ABA, such as GC-ECD or GC-MS. By chromatographing the ABA-ME derivative, the authors purified the extract not only from interfering compounds, but also from possible contaminants in the diazomethane.

1.4.5 Chromatography of ABA in Conjunction with Other Compounds

The general scheme covers the simplest situation, where ABA is separated from all other compounds in an extract. In case where the goal of the experiment is to separate other compounds in conjunction with ABA, the scheme can be modified to include these. The basic approach still holds, however, except that now the operator must simultaneously optimize the system for two or more compounds. Extraction and preparation procedures must be modified to accommodate all compounds of interest (see Vaughan and Milborrow 1984). Gradient elution on a semi-preparative RPLC column will be the likely choice for the first step, followed usually by individual analysis of the compounds on an analytical column of different packing. The recommendations for separating ABA in conjunction with other compounds are particular rather than general, and so the writer refers to the following studies: (1) ABA and its metabolites: Daie et al. (1984), Durley et al. (1982), Feldman et al. (1985), Milborrow and Vaughan (1979), Murphy (1984), Setter et al. (1981), Vaughan and Milborrow (1984), Zeevaart (1980, 1983); (2) ABA and other hormones: Arteca et al. (1980), Durley et al. (1982), Hubick et al. (1986), Rodriguez and Sanchez Tames (1985); (3) ABA and Jasmonic acid: Anderson (1985).

1.4.6 Summary of General Separation Scheme

The writer proposes the general scheme as a strategy for separating ABA. While the writer's approach may not suit other researchers, the general scheme does demonstrate the great versatility of HPLC and the many modifications available to the operator. The methods of greatest convenience and widest applicability were chosen. The RPLC step will accommodate nearly all separations; where it will not, the reader is referred to Sect. 1.2.4), (Modes of Chromatography), for other methods.

References

Anderson JM (1985) Simultaneous determination of abscisic acid and jasmonic acid in plant extracts using high-performance liquid chromatography. J Chromatogr 330:347–355

Arteca RN, Poovaiah BW, Smith OE (1980) Use of high performance liquid chromatography for the determination of endogenous hormone levels in *Solanum tuberosum L.* subjected to carbon dioxide enrichment of the root zone. Plant Physiol 65:1216–1219

Brenner ML (1981) Modern methods for plant growth substance analysis. Annu Rev Plant Physiol 32:511–538

Brown BH, Neill SJ, Horgan (1986) Partial isotope fractionation during high performance liquid chromatography of dueterium-labelled internal standards in plant hormone analysis: a cautionary note. Planta 167:421–423

Cargile NL, Borchert R, McChesney JD (1979) Analysis of abscisic acid by high-performance liquid chromatography. Anal Biochem 97:331–339

Carnes MG, Brenner ML, Andersen CR (1974) Rapid separation and identification of plant growth substances using high pressure liquid chromatography. In: Tamura S (ed) Plant growth substances. Hirokowa, Tokyo, p 99

Chia AJ, Brenner ML, Brun WA (1977) Rapid separation and quantification of abscisic acid from plant tissues using high performance liquid chromatography. Plant Physiol 59:821–826

Creelman RA, Zeevaart JAD (1984) Incorporation of oxygen into abscisic acid and phaseic acid from molecular oxygen. Plant Physiol 75:166–169

Crozier A, Reeve DR (1977) The application of high performance liquid chromatography to the analysis of plant hormones. In: Pilet PE (ed) Plant growth regulation, proceedings of the ninth international conference on plant growth substances. Springer, Berlin Heidelberg New York, p 67

Crozier A, Zaerr JB, Morris RO (1980) High-performance steric exclusion chromatography of plant hormones. J Chromatogr 198:57–63

Daie J, Wyse R, Hein M, Brenner ML (1984) Abscisic acid metabolism by sources and sink tissues of sugar beet. Plant Physiol 74:810–814

During H (1977) Analysis of abscisic acid and indole-3-acetic acid from fruits of *Vitis vinifera L.* by high pressure liquid chromatography. Experimentia 33:1666–1667

Düring H, Bachmann O (1975) Abscisic acid analysis in *Vitis vinifera* in the period of endogenous bud dormancy by high pressure liquid chromatography. Physiol Plant 34:201–203

Durley RC, Kannangara T, Simpson GM (1978) Analysis of abscisins and 3-indolylacetic acid in leaves of *Sorghum bicolor* by high performance liquid chromatography. Can J Bot 56:157–161

Durley RC, Kannangara T, Simpson GM (1982) Leaf analysis for abscisic, phaseic and indolylacetic acids by high-performance liquid chromatography. J Chromatogr 236:181–188

Engelhardt H (1979) High performance liquid chromatography. Springer, Berlin Heidelberg New York

Feldman LJ, Arroyave NJ, Sun PS (1985) Abscisic acid, xanthoxin and violaxanthin in the caps of gravistimulated maize roots. Planta 166:483–489

Fiscus EL, Stermitz FR, Amoros LG (1982) Abscisic acid transport coefficients of *Phaseolus* root systems. Plant Physiol 69:1353–1355

Gaff DF, Loveys BR (1984) Abscisic acid content and effects during dehydration of detached leaves of dessication tolerant plants. J Exp Bot 35:1350–1358

Games DE (1983) High-performance liquid chromatography/mass spectrometry (HPLC/MS). In: Giddings JC, Grushka E, Cazes J, Brown PR (eds) Advances in chromatography, vol 21. Marcel Dekker, New York, p 1

Hubick KT, Reid DM (1980) A rapid method for the extraction and analysis of abscisic acid from plant tissues. Plant Physiol 65:523–525

Hubick KT, Reid DM (1984) Isolation and analysis of plant growth regulators. In: Cell culture and somatic cell genetics of plants, vol 1. Academic Press, New York, p 659
Hubick KT, Taylor JS, Reid DM (1986) The effects of drought on levels of abscisic acid, cytokinins, gibberellins and ethylene in aeroponically-grown sunflower plants. Plant Growth Regul 4:139–151
Jensen E, Rivier L, Junttila O, Crozier A (1986) Identification of abscisic acid from shoots of *Salix pentandra*. Physiol Plant 66:406–408
Jones MG, Metzger JD, Zeevaart JAD (1980) Fractionation of gibberellins in plant extracts by reverse phase high performance liquid chromatography. Plant Physiol 65:218–221
Knox JH, Done JN, Fell AF, Gilbert MT, Pryde A, Wall RA (1978) In: Knox JH (ed) High-performance liquid chromatography. Edinburgh University Press, Edinburgh
Koshioka M, Takeno K, Beal FD, Pharis RP (1983) Purification and separation of plant gibberellins from their precursors and glucosyl conjugates. Plant Physiol 73:398–406
Krstulovic AM, Brown PR (1982) Reversed-phase high-performance liquid chromatography, theory, practice, and biomedical applications. Wiley, New York
Lawrence JF (1981) Organic trace analysis by liquid chromatography. Academic Press, New York
Lewis RW, Visscher SN (1982) A simplified method for the analysis of abscisic acid. Plant Growth Regul 1:25–30
Milborrow BV, Mallaby R (1975) Occurrence of methyl (+)-abscisate as an artefact of extraction. J Exper Bot 26:741–748
Milborrow BV, Vaughan G (1979) The long term metabolism of (±)-[2-^{14}C]abscisic acid by apple seeds. J Exper Bot 30:983–995
Moore R, Smith JD (1984) Growth, graviresponsiveness and abscisic-acid content of *Zea mays* seedlings treated with fluridone. Planta 162:342–344
Murphy GJP (1984) Metabolism of R,S-[2-^{14}C]abscisic acid by non-stressed and water-stressed detached leaves of wheat (*Triticum aestivum* L.). Planta 160:250–255
Ogan K (1984) Detectors for use with small bore columns. In: Scott RP (ed) Small bore liquid chromatography columns: their uses and properties. Wiley, New York, p 85
Quebedeaux B, Sweetser PB, Rowell JC (1976) Abscisic acid levels in soybean reproductive structures during development. Plant Physiol 58:363–366
Reeve DR, Crozier A (1980) Quantitative analysis of plant hormones. In: MacMillan J (ed) Encyclopedia of plant physiology N.S., vol 9; Hormonal regulation of development: molecular aspects. Springer, Berlin Heidelberg New York, p 202
Reeve DR, Crozier A (1983) A reply to 'information theory and plant growth substance analysis' by I. M. Scott. Plant Cell Environ 6:365–368
Robertson JM, Pharis RP, Huang YY, Reid DM, Yueng EC (1985) Drought-induced increases in abscisic acid levels in the root apex of sunflower. Plant Physiol 79:1086–1089
Rodriguez A, Saanchez Tames R (1985) Analysis of 3-Indolylacetic acid and abscisic acid by high-performance liquid chromatography and gas-liquid chromatography. Anal Biochem 146:184–190
Saunders PF (1978) The identification and quantitative analysis of abscisic acid in plant extracts. In: Hillman JR (ed) Isolation of plant growth substances, society for experimental biology seminar series. Cambridge University Press, Cambridge, p 115
Scott IM (1982) Information theory and plant growth analysis. Plant Cell Environ 5:339–342
Scott IM (1983) An answer to Reeve & Crozier. Plant Cell Environ 6:367–368
Setter TL, Brun WA, Brenner ML (1981) Abscisic acid translocation and metabolism in soybeans following depodding and petiole girdling treatments. Plant Physiol 67:774–779
Simpson CF (ed) (1976) Practical high performance liquid chromatography. Heyden and Son, London
Snyder LR, Kirkland JJ (1979) Introduction to modern liquid chromatography, 2nd edition. Wiley, New York

Snyder LR (1980) Gradient elution. In: Horvath C (ed) High performance liquid chromatography, advances and perspectives, vol 1. Academic Press, New York, p 208

Sweetser PB, Vatvars A (1976) High-performance liquid chromatographic analysis of abscisic acid in plant extracts. Anal Biochem 71:68–78

Taylor JS, Reid DM, Pharis RP (1981) Mutual antagonism of sulfur dioxide and abscisic acid in their effect on stomatal aperture in broad bean (*Vicia faba L.*) epidermal strips. Plant Physiol 68:1504–1507

Vaughan GT, Milborrow BV (1984) Resolution of RS-abscisic acid and the separation of abscisic acid metabolites from plant tissue by high-performance liquid chromatography. J Chromatogr 336:221–228

Yokota T, Murofushi N, Takahashi N (1980) Extraction, purification, and identification. In: MacMillan J (ed) Encyclopedia of plant physiology N.S., vol 9. Hormonal regulation of development: Molecular aspects. Springer, Berlin Heidelberg New York, p 111

Zeevaart JAD (1980) Changes in the levels of abscisic acid and its metabolites in excised leaf blades of *Xanthium strumarium* during and after water stress. Plant Physiol 66:672–678

Zeevaart JAD (1983) Metabolism of abscisic acid and its regulation in *Xanthium* leaves during and after water stress. Plant Physiol 71:477–481

High Performance Liquid Chromatography and the Analysis of Indole-3-Acetic Acid, and Some of Its Decarboxylated Catabolites in Scots Pine (*Pinus sylvestris* L.)

G. SANDBERG, A. CROZIER, A. ERNSTSEN, and B. SUNDBERG

1 Introduction

Indole-3-acetic acid (**1**) (IAA) is a weak organic acid with a pK_a of 4.8 and is an important growth-regulating substance in higher plants (see Letham et al. 1978). The biological activity of submicrogram amounts of IAA has been demonstrated by exogenous applications to several different species and tissues. It is believed that endogenous IAA is involved in the regulation of plant growth and development, controlling processes such as cell elongation and cell differentiation. The first conclusive evidence of the presence of IAA in higher plants, was obtained with *Zea mays* by Haagen-Smit et al. (1946). Since then IAA has been identified as an endogenous constituent of a number of plant species and its function as a growth regulator is thought to be universal. The concentration of IAA in plant tissues is usually around 1–100 ng g^{-1} fresh weight. When investigating the hormonal role of endogenous IAA, it is common practice to attempt to correlate changes in IAA levels with the developmental event under study. However, this approach will be of limited value if the amount of IAA available for growth is related to the rate of turnover rather than to the pool size. It is therefore essential to investigate IAA metabolism pathways and to monitor not only IAA pools but also those of its precursors and catabolites.

The catabolism of IAA is still a controversial topic despite having been investigated for more than three decades. Two degradative routes are reported to operate, namely a decarboxylative and a nondecarboxylative pathway (Sembdner et al. 1980). Although many of the components in the nondecarboxylative pathway have been detected in extracts from higher plants, experimental evidence on the exact catabolic routes operating in individual species is very limited indeed (Bandurski and Nonhebel 1984). Decarboxylative catabolism of IAA, which involves a branched pathway leading to either 3-hydroxymethyloxindole, 3-methyleneoxindole and 3-methyloxindole, or alternatively indole-3-methanol (**2**) (IM), indole-3-aldehyde (**3**) (IAld), and indole-3-carboxylic acid (**4**) (ICA), has been investigated in vitro using assorted IAA oxidase preparations and commercially available enzymes, such as horseradish peroxidase (HRP), and in vivo using excised tissues and whole plant test systems (see Sembdner et al. 1980). Unfortunately, characterization of the catabolites of IAA in most of these studies has been somewhat tenuous and few of the compounds potentially involved have been identified on the basis of strict physicochemical evidence. Moreover, only IAld, ICA, and IM have been identified as endogenous constituents of higher plants

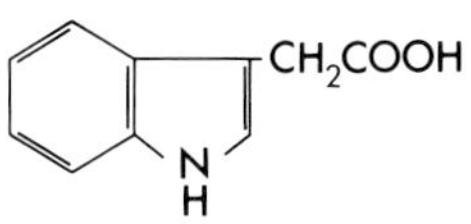

indole-3-acetic acid (1)

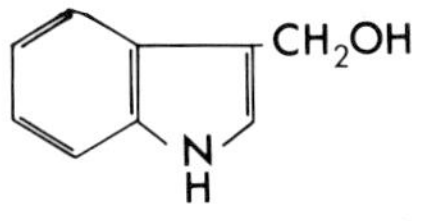

indole-3-methanol (2)

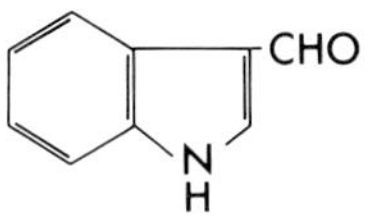

indole-3-aldehyde (3)

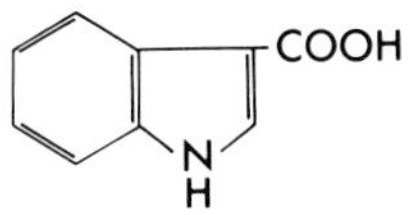

indole-3-carboxylic acid (4)

(Bearder 1980; Sandberg et al. 1984; Sundberg et al. 1985). In this review we report on the use of high performance liquid chromatography (HPLC) and ancillary purification procedures employed in the analysis of IAA and some of its decarboxylated catabolites in Scots pine (*Pinus sylvestris* L.).

In the 1960's the analysis of IAA and other indoles in plant extracts was usually undertaken with either paper or thin layer chromatography (TLC) combined with either bioassay or colorimetric detection after addition of a chromatogenic agent. Many inventive bioassays for the quantitative detection of auxins were developed during the early years of hormone research (see Reeve and Crozier 1980). The most commonly used are based on the elongation or curvature responses of oat, pea, and wheat coleoptile tissues. The sensitivity of several of these assays is acceptable, but the selectivity is too low and the response is adversely affected by impurities in plant extracts (Reeve and Crozier 1980). Colorimetry of indoles has been widely used in conjunction with paper chromatography and TLC (Ehmann 1977). Such methods normally demand extensive sample purification and are rather insensitive. Not surprisingly, these procedures have now been superseded and present-day analyses usually involve the use of either gas chromatography (GC), high performance liquid chromatography (HPLC) or immunoassay-based methodology.

GC is a very useful procedure for the quantification of indole auxins. This is due mainly to the availability of high efficiency capillary columns and selective detectors that offer very low limits of detection. For instance, electron capture detection of trifluoroacetyl IAA methyl ester and heptafluorobutyryl IAA methyl ester facilitates a detection limit of ca. 10 pg (Seeley and Powell 1974). Similar sensitivity coupled with high selectivity is obtained when methylated indoles are analyzed by GC using a nitrogen-phosphorous detector (Cohen and Schulze 1981; Martin et al. 1980). GC can also be used for combined GC-mass spectrometry (GC-MS). GC-MS is a most powerful and flexible analytical tool and when used in the selected ion current monitoring mode, provides both selectivity and high sensitivity (Rivier and Pilet 1974). Unfortunately, despite recent reductions in the price of compact, bench-top quadrupole instruments, GC-MS is beyond the budget of most laboratories.

The quantification of low concentrations of IAA can also be achieved by immunological assays. Radioimmunoassays (RIA) (Pengelly and Meins 1977; Weiler 1981) and a solid-phase enzyme immunoassay (Weiler et al. 1981) for IAA have been described. Immunological assays are sensitive, selective, do not require prohibitively expensive equipment and are easily adapted to routine analyses. Purification of the sample cannot, however, be excluded, at least when the method is applied to conifer tissues (Sandberg et al. 1985). Preparation of antibodies requires the synthesis of an IAA conjugate, usually IAA-bovine serum albumin, and the subsequent production of antiserum. Relatively large amounts of standard compounds are required for the immunization, and radiolabeled standards of high specific activity are needed for use as RIA immunotracers. Immunological techniques for the analysis of indoles other than IAA have yet to be described.

HPLC was introduced into plant hormone analysis in the early 1970's. Since then reports dealing with the development and adaptation of HPLC methods for indole analysis have appeared with increasing frequency. HPLC has a relatively high sample capacity and, unlike GC, operates at ambient temperature. In addition, samples can be analyzed in either a derivatized or nonderivatized form and monitored with nondestructive detector systems. As a consequence HPLC is readily utilized as either a purification step or an analytical procedure. A further advantage with HPLC is that a wide variety of columns can be used, each providing a different separatory mechanism.

2 Internal Standardization

Quantification of plant hormones generally involves a number of purification steps. These procedures inevitably result in sample losses which adversely affect the reliability of the final quantitative estimate. Sample losses can, however, be accounted for when an isotopically labeled analog of the compound under study is used as an internal standard. The validity of the estimate obtained from a quantification is dependent on the radiochemical purity of the internal standard added to the extract, and on the purity of the labeled material subjected to radiocounting after the purification process. The radiochemical purity of the internal standard must, therefore, be checked prior to addition to the extraction medium. The purity of the recovered material subjected to the radioactivity determination in quantitative HPLC is dependent upon the separatory capacity of the HPLC system, and it is important that losses are estimated after the HPLC separation, and not in an aliquot of the sample prior to HPLC. This is of particular importance when working with indoles which are relatively unstable both in standard solutions and in environments encountered frequently during extraction and purification.

IAA (**1**) and some of its metabolites, labeled in several positions with ^{14}C, ^{3}H, or ^{2}H, can be purchased from commercial sources. Labeling with the radioactive isotopes ^{14}C or ^{3}H must be used whenever possible when estimating recoveries of a compound during a purification procedure leading to quantitative analysis by HPLC. [1-^{14}C]- and [2-^{14}C]IAA are available commercially from several sources. [^{3}H]-labeled IAA is more difficult to purchase due to poor stability during synthesis and storage. [^{3}H]IAA can, however, be synthesized and purified in several different ways. Melhado et al. (1982) described a catalytic reduction to produce [5-^{3}H]IAA. 5-Bromo-IAA (0.168 g) dissolved in 15 ml ethyl acetate was mixed with 0.1 g 5% palladium on charcoal in the presence of 75 mg KOH in 10 ml water, placed in a round-bottomed flask attached to a vacuum line, frozen, and evacuated. Tritium gas was introduced into the flask until atmospheric pressure was obtained and the reaction mixture was stirred at room temperature for 2 h. The solution was filtered, adjusted to pH 11.0 with KOH and washed twice with 25 ml diethyl ether. The aqueous phase was adjusted to pH 3.0 and partitioned four times against 25 ml diethylether. The tritiated IAA in the acidic ether fraction had a specific activity of 61.4×10^{10} Bq mmol^{-1} and was stable for several months when stored in ethanol at −20 °C. The stability of [5-^{3}H]IAA may be related to the purity of the stock solution. Purification by both reversed- and normal phase HPLC before storage is therefore recommended in order to extend shelf-life.

Labeled IM (**2**) and IAld (**3**) can be produced from isotopically labeled IAA by enzymatic conversion with HRP (Ernstsen and Sandberg, unpublished results). [2-^{14}C]IAA (20 μM) is dissolved in 15–30 ml 0.05 M, pH 5.3 phosphate buffer containing 100 nM HRP (Boehringer, Mannheim, West Germany) and stirred for 20 min at room temperature. After adjusting the pH to 7.5, the solution is applied to an XAD-7 column equilibrated with 0.1 M, pH 7.5 phosphate buffer (see Sect. 5.1.2.1). The column is washed with approximately 50 ml of the buffer, containing 0.02% of the antioxidant sodium diethyldithiocarbamate, then eluted with 20 ml ethyl acetate, which also contains 0.02% of the antioxidant. The ethyl acetate fraction is reduced to dryness and analyzed by ion-pair reversed phase HPLC using a radioactivity monitor (Fig. 1) (see Sect. 4). The IM and IAld peaks are collected, and after evaporation of the methanol, IM and IAld are partitioned into ether and the organic phases reduced to dryness. The identity of the products can be confirmed by GC-MS and the isotope contents determined from the mass spectra by the method of Bowen et al. (1972). Specific activities can also be

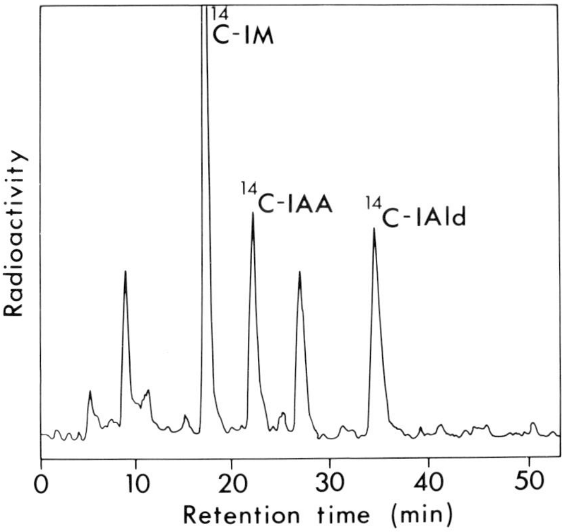

Fig. 1. Reversed phase ion-pair HPLC of indole-3-acetic acid and products formed during incubation of indole-3-acetic acid with horseradish peroxidase. Column: 250 × 4 mm i.d. 5 μm LiChrosorb RP-18. Mobile phase: 20 mM tetrabutylammonium hydrogen sulphate in 50 mM, pH 6.5 sodium phosphate buffer/methanol (65:35, v/v). Flow rate: 0.6 ml min^{-1}. Detector: Reeve Analytical radioactivity monitor, heterogeneous mode, 170 μl flow cell packed with a lithium glass scintillator

checked by quantitative HPLC analysis and liquid scintillation counting. No significant differences between the specific activities of the [2-^{14}C]IAA fed and the recovered products have been observed in our laboratory.

3 Extraction and Purification

The efficiency of the extraction process has not been thoroughly investigated, despite its importance in quantitative analysis of phytohormones. Analytical techniques for growth-regulating substances commonly give a precision of 10 to 30% in post-extraction sections of a purification procedure. However, no data are available on the efficiency of the extraction process itself. Thus, errors associated with the actual removal of the compounds of interest from the tissue may be considerably larger than those associated with sample purification.

Methanol and, to some extent, ethanol are the most commonly used extractants for indoles. These alcohols extract pigments from the tissue, which indicates plastid breakdown and general denaturation of membranes. Aqueous methanol is also known to inhibit enzyme activity (Libbert et al. 1969). One potential drawback when using either methanol or ethanol as extractants is the risk of partial esterification of carboxylic acids to their respective methyl and ethyl esters (Allen et al. 1982). It is our experience, however, that this is not a significant problem. Diethyl ether and acetone have also been used for the extraction of indoles from plant tissues. Hemberg and Tillberg (1980) reported that the amount of IAA extracted from dry *Zea mays* kernels with diethyl ether increases continually over a 40-day extraction period. While this could be due to inefficient short-term extraction of IAA, it is more likely to be a consequence of the progressive hydrolysis of IAA conjugates (Srivastava 1963) coupled with the possible formation of IAA through breakdown of other indoles. Extraction of *Zea* kernels with absolute acetone gives a low level of IAA which does not increase significantly with either prolonged extraction periods or increased temperatures. However, a dramatic increase in IAA yield occurs when small amounts of water are added to the absolute acetone (Hemberg and Tillberg 1980).

Extraction with 0.1 M, pH 8.0 phosphate buffer has been used by Sundberg et al. (1985) as a means of reducing nonpolar contaminants such as pigments and lipids. The major disadvantage of this approach is the possibility of enzymatic conversions during extraction. However, several precautions can be taken to reduce this risk. They include protein precipitation, ultrasonication, and working at low temperatures. It should also be noted that enzymatic degradation of IAA in a crude enzyme extract from *Pinus sylvestris* has been shown to be optimal at pH 4–5. Thus, under mild basic conditions breakdown of IAA is likely to be negligible.

Nonenzymatic oxidation of IAA occurs in the presence of light, oxygen, or peroxide, and it is also degraded by acids (Sembdner et al. 1980). Degradation thus occurs only too readily in conditions that prevail during the extraction and purification of endogenous IAA. This also applies to several other indoles. For instance, the biosynthetic intermediate, indole-3-pyruvic acid, is reported to be converted to IAA, as well as to other indoles, during sample purification (Atsumi et al. 1976; Sweetser and Swartzfager 1978; Hemberg and Tillberg 1980). However, low temperatures and the use of an antioxidant in the extraction medium reduced greatly the decomposition of indole-3-pyruvic acid (Iino and Carr 1982). If an antioxidant, such as butylated hydroxytoluene or sodium diethyldithiocarbamate, is added to all solvents from the extraction medium through to the final HPLC step, it is our experience that degradation of indoles in general is minimal.

Historically, partitioning between two immiscible liquids has been used widely to purify and fractionate plant extracts. Several different solvents have been used over the years. In practice, diethyl ether or ethyl acetate are usable for most purification and/or fractionation procedures involving nonconjugated indoles. Recently, a procedure based on Amberlite XAD-7 was introduced as an alternative to partitioning procedures (Andersson and Andersson 1982) and the procedure has been used for the purification of several indoles, including ICA and IM (Sandberg et al. 1984; Sundberg et al. 1985). Practical details of this technique are presented in Sect. 5.1.2.1.

Purification of indoles in plant extracts can be achieved by column chromatography using supports such as polyvinyl-N-pyrrolidone (PVP) and/or Sephadex LH-20 (Sandberg et al. 1981 b). These materials have the advantage of drastically reducing the dry weight of the sample and at the same time giving high recoveries of indolic compounds. Several types of commercially available cartridges prepacked with silica or silica-bonded reversed phase materials can also be used as rapid purification and/or concentration steps. Although as yet its use has been restricted to IAA, immunoaffinity chromatography is an extremely efficient purification step and it must be regarded as one of the few techniques currently available with the potential for processing large numbers of samples (Sundberg et al. 1986).

4 High Performance Liquid Chromatography

A variety of HPLC systems can be employed to analyze indoles although, in practice, usage has been restricted largely to ion-suppression and ion-pair reversed phase and normal phase separations. Detailed information on ion-suppression reversed phase HPLC of indoles can be obtained from Jensen (1982) and Wurst et al. (1980, 1984). Jensen (1982) investigated the influences of the organic modifier, pH of the mobile phase and stationary phase alkyl chain length. Analysis of certain indoles is complicated by their instability. IM, for instance, breaks down very rapidly under acidic conditions (Sundberg et al. 1985), thus a neutral aqueous phase must be employed when it is analyzed by reversed phase HPLC. Similarly, when IM is analyzed by normal phase HPLC, the trace of acetic acid that is usually added to the mobile phase to suppress tailing of acidic indoles must be omitted. Sandberg et al. (1981 a) have published details of the factors controlling

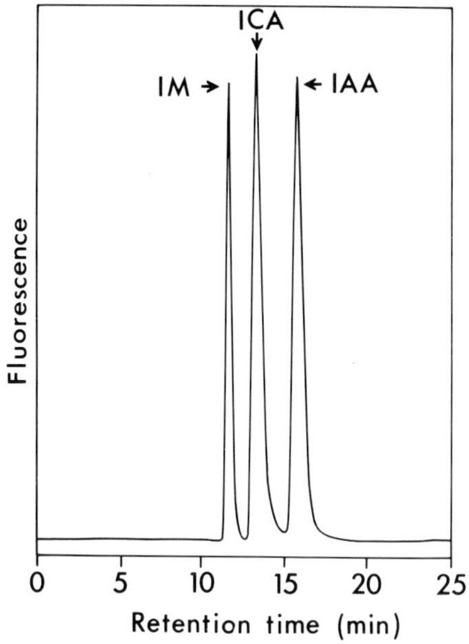

Fig. 2. Reversed phase ion-pair HPLC of indole-3-methanol, indole-3-carboxylic acid and indole-3-acetic acid. Column: 250 × 4 mm i.d. 5 μm LiChrosorb RP-18. Mobile phase: 20 mM tetrabutylammonium hydrogen sulphate in 50 mM, pH 6.5 sodium phosphate buffer/methanol (67:33, v/v). Flow rate: 1 ml min^{-1}

ion-pair HPLC separations of indoles. Ion-pair HPLC is especially useful for the separation of IAA and ICA (see Fig. 2) as these compounds exhibit very similar retention properties in many ion-suppression and normal phase systems.

Detection of indoles in the HPLC eluent can be accomplished with an absorbance monitor, an electrochemical detector or a fluorimeter. An absorbance monitor operating at 280 nm provides nonselective detection in the nanogram range. An electrochemical detector is selective, in that many extract impurities will not induce a significant response, and is sensitive to ca. 200 pg of IAA. Its use, for practical intents and purposes, is restricted to aqueous-based solvent systems and regular maintenance of the electrode is essential. A fluorimeter is likely to be a much more reliable proposition for routine selective detection of indoles. Although individual fluorescence λ_{max} will vary slightly (see Burnett and Audus 1964), most indoles [IAld (**3**) is one exception] can be readily detected with an instrument operating at excitation 280 nm and emission 350 nm. The limit of detection is very much dependent upon the type of fluorimeter used. A good quality instrument can detect as little as 1 pg of IAA while a cheap filter detector may be able to detect only ca. 1 ng. Recently, an HPLC system based on precolumn derivatization and fluoresence detection of the indolo-α-pyrone derivative of IAA (Stoessl and Venis 1970) has been described by Blakesley et al. (1983). When analyzing radioactive samples by HPLC an on-line radioactivity monitor can be used as a detector system (Reeve and Crozier 1977, 1983) (see Fig. 1).

5 Examples

5.1 Indole-3-Acetic Acid

Numerous purification procedures for the analysis of IAA (**1**) in plant tissues have been used over the years. Only a few, however, have been adapted for use with conifer tissues (De Yoe and Zaerr 1976; Caruso et al. 1978; Crozier et al. 1980; Sandberg et al. 1981 a; Caruso and Zeisler 1983; Blakesley et al. 1983). In recent years, attempts have been made to modify time-consuming, but accurate techniques in order to facilitate rapid multisample handling. More often than not this produces a conflict between speed of analysis and accuracy. This was highlighted with the introduction of immunological assays for IAA analysis. The selectivity of these techniques has been claimed to be high enough to permit quantification of IAA in unpurified methanolic extracts (Weiler 1982). Tests with conifer extracts, however, have shown that extensive sample purification is essential if accurate estimates are to be obtained with RIA (Sandberg et al. 1985).

5.1.1 Extraction

Tissues of *Pinus sylvestris* are homogenized and extracted for 2 h in methanol (0.3 g ml^{-1}) containing 0.02% sodium diethyldithiocarbamate and [1-^{14}C]IAA as an internal standard. After filtration the methanolic extract is reduced to dryness in vacuo at 40 °C, dissolved in 20 mM, pH 7.5 phosphate buffer (2.5 ml g^{-1} tissue), and filtered to remove particulate material prior to further analysis.

5.1.2 Purification

Two different procedures are described for the purification of IAA in pine extracts (Fig. 3). The first is more traditional, utilizing PVP and XAD chromatog-

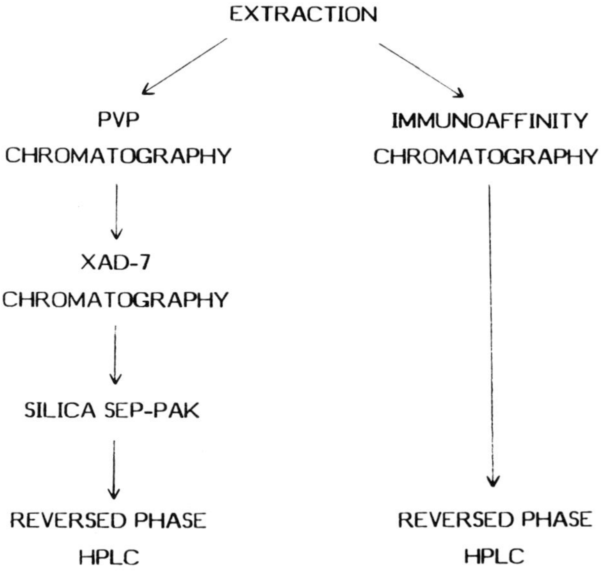

Fig. 3. Procedures for quantitative analysis of indole-3-acetic acid

raphy. The second uses immunoaffinity column as the sole purification step (Sundberg et al. 1986).

5.1.2.1 PVP, XAD-7, and Sep-Pak

The buffer phase obtained after extraction is applied to a 100 × 10 mm i.d. column packed with PVP. Technical grade PVP is normally purchased and before use is repeatedly slurried in distilled water to remove very small particles. The fined PVP is packed into the column by gravity sedimentation and prior to use is eluted with at least five volumes of buffer. The aqueous sample at pH 7.5 is applied to the column which is eluted with 20 ml 20 mM, pH 7.5 phosphate buffer. The eluent from the PVP column is adjusted to pH 2.7 and applied to a 150 × 5 mm i.d. column packed with XAD-7. Analytical grade XAD-7 is rather expensive, and it may be advantageous to purchase the technical grade and before use wash up to five times in a solvent series comprised of ethyl acetate, methanol, and water. In practice, the XAD-7 resin (0.3 g dry weight for medium-sized samples) can be packed in a 2-ml glass pipette with a glass wool plug at the base. The XAD column is equilibrated with 10 ml of buffer and after the sample is then run through it, is washed with 5 ml of n-hexane to remove excess water and non-polar contaminants. At this juncture a silica Sep-Pak (Waters Associates, Milford, Massachusetts, USA) is attached to the outlet of the XAD column. The two columns are then eluted in series with 20 ml of dichloromethane. IAA is now eluted from the XAD and adsorbed on to the Sep-Pak. The Sep-Pak is disconnected from the XAD column, washed with 5 ml of ethyl acetate/hexane (3:1, v/v) and the eluate, which contains IAA, collected and reduced to dryness prior to analysis by HPLC.

5.1.2.2 Immunoaffinity Chromatography

This procedure makes use of IAA antibodies raised against IAA bovine serum albumin coupled via the nitrogen of the indole ring (Pengelly and Meins 1977). The antibodies are bound to a glutardialdehyde-activated affinity adsorbent which is then used as a support for immunochromatography of IAA (Sundberg et al. 1986).

a) Preparation and Purification of Antiserum. Rabbits, 3–4 months old, are immunized with an emulsion of IAA-bovine serum albumin conjugate, coupled to the nitrogen of the indole ring, in buffered physiological saline solution and Freunds incomplete adjuvant. Blood is collected at monthly intervals and the serum stored at −20 °C or below. Purification of the immunogammaglobulin (IgG) fraction is carried out on a 50 × 10 mm i.d. column packed with protein A-sepharose CL-48 (Pharmacia, Uppsala, Sweden). The serum is mixed 1:1 (v/v) with 0.1 M, pH 7.0 phosphate buffer and applied to the column which is then eluted with the buffer until no more UV absorbing impurities are removed. The IgG fraction is eluted with 0.1 M, pH 3.0 citrate buffer. The pH is adjusted to 7.0 with 1.0 N NaOH and the volume is reduced to the original serum volume by membrane ultrafiltration. The amount of IgG, *C*, can be calculated by measuring the absorbance at 280 nm, using the formula,

$$C = \frac{A_{280}}{1.4} \text{ mg ml}^{-1} .$$

b) Preparation of the Immunoaffinity Support. Glutardialdehyde-activated affinity adsorbent (Boehringer, Mannheim, West Germany) is suspended in a 0.9% NaCl solution containing the IgG (1 g adsorbent, 50 mg IgG/5 ml). The mixture is degassed in vacuo and shaken gently for 4 h at room temperature before being transferred to a 10 mm i.d. chromatography column and washed with a 1.5% solution of NaCl until the eluate is free of protein (A_{280} <0,005). Free aldehyde groups remaining on the adsorbent are thereafter charged by incubation with ethanolamine.

c) Immunoaffinity Chromatography of Extracts. An extract from 1.0 g of tissue prepared as described above (Sect. 5.1.1), is dissolved in 2.5 ml of a 20 mM, pH 7.5 phosphate buffer and applied to a 35 × 10 mm i.d. IgG affinity column (ca. 1.7 g absorbent) which is eluted for 25 min at a flow rate of 2 ml min^{-1}. The IAA fraction is then eluted from the column with 30 ml of 0.1 M, pH 2.7 citrate buffer which is either partitioned three times against half volumes of diethyl ether or concentrated on a XAD-7 column. Before being re-used, the immunoaffinity column must be cleaned with 200 ml of 1 M propionic acid and re-equilibrated in pH 7.5 phosphate buffer.

5.1.3 HPLC Analysis

Either ion-suppression or ion-pair reversed phase HPLC with fluorescence detection are suitable for quantitative analysis of IAA in *Pinus sylvestris* needle extracts following purification by either of the procedures described in Sect. 5.1.2.1 and 5.1.2.2. Purification by immunoaffinity chromatography is both rapid and extremely effective (Fig. 4). Irrespective of the HPLC system used, the amount of unlabeled IAA present in the purified extract is estimated from the size of the IAA-like fluorescent peak while the amount of radioactivity in this zone enables sample losses to be determined. Quantitative estimates of endogenous IAA in the original methanolic extract (*X*) are calculated according to the following formula

$$X = \frac{(\text{IAA detected})\ ([^{14}\text{C}]\text{IAA added})}{([^{14}\text{C}]\text{IAA detected})} - [(\text{sp. act.}\ [^{14}\text{C}]\text{IAA})\ ([^{14}\text{C}]\text{IAA added})] . \quad (1)$$

If such estimates are to be accurate, it is imperative that impurities do not make a significant contribution to the putative IAA peak. Thus before routine analysis of extracts from a particular tissue is attempted, preliminary experiments should be initiated to test the homogeneity of the HPLC peak used to quantify IAA. This can easily be achieved by either analyzing the putative IAA peak in a normal phase HPLC system (Fig. 5) or subjecting it to methylation prior to reanalysis by reversed phase HPLC (Fig. 6). The presence of a single fluorescent

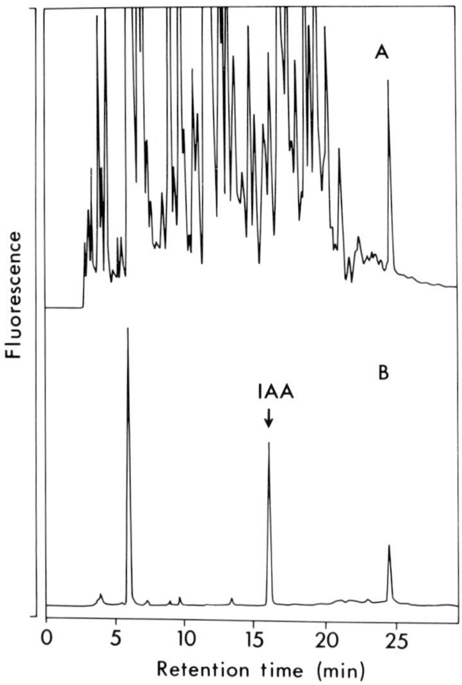

Fig. 4. Reversed phase ion-suppression HPLC of extract from *Pinus sylvestris* before and after purification on an indole-3-acetic acid-immunoaffinity column. Column: 250 × 4 mm i.d. 5 µm Hypersil ODS. Mobile phase: 25 min gradient 25–75% methanol in 1% aqueous acetic acid. Flow rate: 1 ml min^{-1}

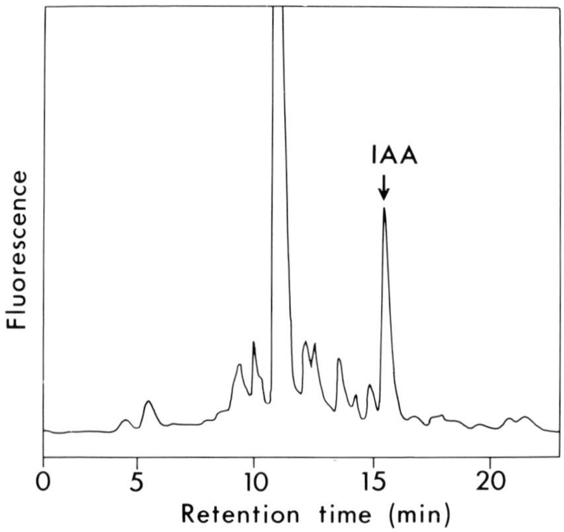

Fig. 5. Reversed phase ion-suppression HPLC of semi-purified extract from *Pinus sylvestris. Arrow* indicates peak co-chromatographing with indole-3-acetic acid. Column: 250 × 4 mm i.d. 5 µm LiChrosorb RP-18. Mobile phase: 40% methanol in 1% aqueous acetic acid. Flow rate: 1 ml min^{-1}

peak at the correct retention time, such as illustrated in Fig. 6, provides good evidence of peak homogeneity and hence the accuracy of the original reversed phase analysis. Alternatively, accuracy can be assessed more formally by carrying out a successive approximation as described by Reeve and Crozier (1980). In practice, this involves analyzing and collecting the putative IAA peak from the initial reversed phase separation and reanalyzing it by, say, normal phase HPLC. If the

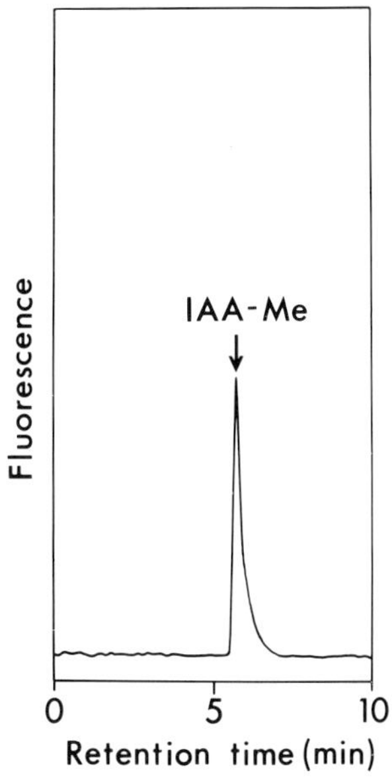

Fig. 6. Reversed phase ion-suppression HPLC of methylated indole-3-acetic acid fraction. Column: 250 × 4 mm i.d. 5 µm LiChrosorb RP-18. Mobile phase: 60% methanol in 1% aqueous acetic acid. Flow rate: 1 ml min^{-1}

second quantitative estimate of IAA, X_2, calculated from Eq. (1), is not significantly different from the first, X_1, there is a sound basis for assuming that X_1 is accurate. If a difference is found, then X_1 must be rejected as inaccurate and the sample further purified until an estimate is obtained that does not change upon purification.

5.2 Indole-3-Carboxylic Acid

ICA (**4**) is present in *Pinus sylvestris* in much lower amounts than IAA (Sandberg et al. 1984). As a consequence relatively large amounts of tissue have to be extracted (ca. 20 g) and samples must be purified extensively prior to analysis by HPLC. Unfortunately, radiolabeled ICA is not available at present for use as an internal standard, thus losses have to be estimated indirectly from recoveries of [^{14}C]IAA. To further compound the situation, it is at least possible that a portion of the ICA detected in an extract could arise from breakdown of IAA during sample purification (Gmelin and Virtanen 1961; Gmelin 1964). It is therefore essential to distinguish between endogenous ICA and ICA arising from degradation of IAA. This, as will be described in Sect. 5.2.3, can be achieved by using [2-^{14}C]IAA as an internal standard and analyzing both IAA and ICA according to the procedures of Sandberg et al. (1984) as outlined in Fig. 7.

5.2.1 Extraction and Purification

Pinus sylvestris needles and a [2-^{14}C]IAA internal standard are homogenized in methanol as described in Sect. 5.1.1 and the methanolic extract reduced to dryness

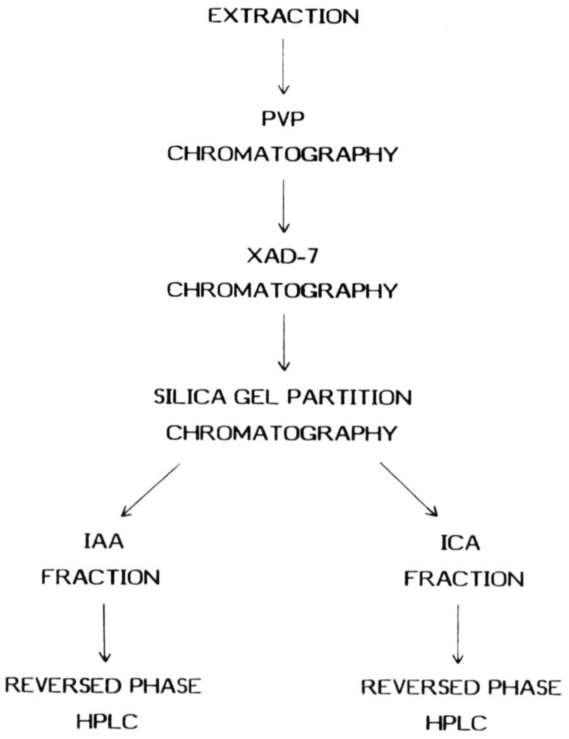

Fig. 7. Procedures for quantitative analysis of indole-3-carboxylic acid

in vacuo at 40 °C. The extract is dissolved in 200 ml 0.1 M, pH 8.0 phosphate buffer and washed five times with petroleum ether. The buffer phase is either slurried with PVP or applied to the top of a 450 × 10 mm i.d. column of PVP which is then eluted with 300 ml 0.1 M, pH 8.0 phosphate buffer. The eluate from the PVP column is adjusted to pH 2.7 and run through a 150 × 10 mm i.d. XAD-7 column which is subsequently eluted with hexane/ethyl acetate (40:60, v/v). ICA and IAA are collected in the 5–30 ml fraction, which is reduced to dryness and purified further on a 500 × 10 mm i.d. column packed with 10 μm silica gel coated with a 40% loading (w/v) of a 0.5 M formic acid stationary phase. With a mobile phase of 0.5 M formic acid saturated-hexane/ethyl acetate (80:20, v/v) IAA elutes between 14.0 and 16.5 ml while ICA has a retention volume of 16.5–18.0 ml. The two zones are collected separately and analyzed by HPLC.

5.2.2 HPLC Analysis

The IAA fraction is analyzed by reversed phase HPLC as described in Sect. 5.1.3 and evidence of peak homogeneity is obtained by further analysis in a normal phase system. A suitable system would be a 250 × 5 mm i.d. column packed with either a 5 μm Hypersil CPS, Nucleosil cyanopropyl or Spherosorb nitrile support, and eluted with a mobile phase of ca. 75:25 (v/v) hexane/ethyl acetate in 1% acetic acid.

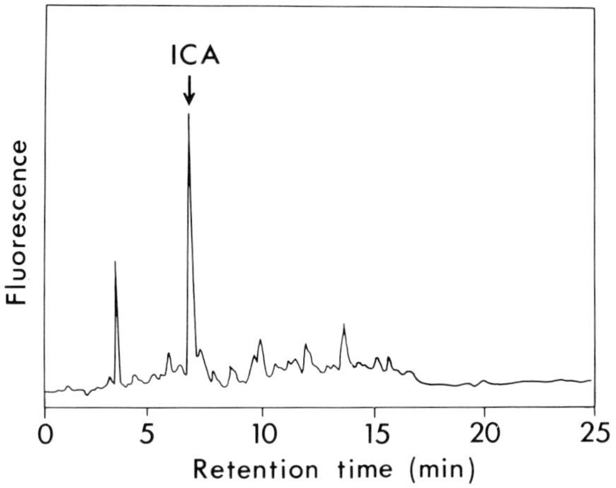

Fig. 8. Reversed phase ion-suppression HPLC of semi-purified extract from *Pinus sylvestris* needles. *Arrow* indicates peak co-chromatographing with indole-3-carboxylic acid. Column: 150 × 4.6 mm i.d. 5 μm Lichrosorb RP-18. Mobile phase: 25 min gradient 20–60% methanol in 1% aqueous acetic acid. Flow rate: 1 ml min^{-1}

ICA, like IAA, can be analyzed by either ion-suppression or ion-pair reversed phase HPLC. A typical trace obtained when the ICA fraction from *Pinus sylvestris* is analyzed by ion-suppression HPLC is illustrated in Fig. 8. Verification of the accuracy of quantitative estimates based on the fluorescent ICA peak can be obtained by a subsequent normal phase HPLC analysis in the system outlined in the previous paragraph.

5.2.3 Internal Standards

In order to assess the possible breakdown of IAA to ICA, 100 g samples of *Pinus sylvestris* needles were extracted along with 25×10^3 Bq [2-^{14}C]IAA (2.6 μg) and 10 μg of ICA as a cold carrier (Sandberg et al. 1984). ICA and IAA were then analyzed according to the procedures outlined above and in Fig. 8. The amounts of unlabeled ICA and IAA remaining at the end of the analytical sequence were estimated from the size of the appropriate fluorescent HPLC peaks, while radiolabeled species were quantified by liquid scintillation counting. These data provide a measure of the percentage breakdown of IAA to ICA (Y) calculated from Eq. (2).

$$Y = \frac{(\text{ICA added})\ ([^{14}\text{C}]\text{ICA detected}) \times 100}{(\text{ICA detected})\ ([^{14}\text{C}]\text{IAA added})}\ \% . \qquad (2)$$

In a second series of experiments, 20 g of *Pinus* needles were analyzed using 1.5×10^3 Bq [2-^{14}C]IAA (156 ng) as an internal standard. Quantitative estimates of endogenous IAA (X) were obtained from Eq. (1) (see Sect. 5.1.3) while endogenous ICA (Z) was calculated according to Eq. (3)

$$Z = \frac{(\text{ICA detected})\ ([^{14}\text{C}]\text{IAA added})}{([^{14}\text{C}]\text{IAA detected})} - [(X + [(\text{sp. act. }[^{14}\text{C}]\text{IAA})\ ([^{14}\text{C}]\text{IAA added})]) \times Y] . \qquad (3)$$

While processing the data in this manner provides a means of estimating both sample losses and the amount of ICA originating from IAA, the relatively low specific activity of the [2-^{14}C]IAA (1.63×10^9 Bq mmol^{-1}) and the lack of an ICA internal standard labeled with a different radioisotope necessitate a number of compromises in experimental design. As a consequence, ICA losses have to be calculated indirectly from [2-^{14}C]IAA recoveries. Furthermore, as values of Y cannot be determined for individual extracts, allowance for this factor has to be based on a mean value obtained in the experiments in which 25×10^3 Bq rather than 1.5×10^3 Bq of [2-^{14}C]IAA was used as an internal standard.

Application of Eq. (2) to the data obtained from analyses of 100 g tissue samples, to which 25×10^3 Bq [2-^{14}C]IAA and 10 μg ICA were added as internal standards, indicated that $0.6 \pm 0.1\%$ ($\pm$SD) of the IAA pool was broken down to ICA during purification. When the data from analyses of 20 g samples of *Pinus* needles and 1.5×10^3 Bq [2-^{14}C)IAA were processed by Eq. (1), an estimate of 25.4 ± 6.5 ng IAA g^{-1} ($\pm$SD) was obtained. Equation (3) provided an estimate of 2.5 ± 0.4 ng ICA g^{-1}, of which 2.3 ng was calculated to represent endogenous ICA and 0.2 ng assessed as having originated from the breakdown of both endogenous IAA and IAA added to extracts as an internal standard.

5.3 Indole-3-Methanol

IM (**2**) is a neutral indole which decays rapidly in an acidic environment (Sundberg et al. 1985). As a consequence it is essential that low pH's are avoided during both purification and analysis. Because of the action of peroxidases, it is possible that IM could be formed from IAA during sample preparation. Attempts must therefore be made to minimize peroxidase activity and the extent of possible conversions of IAA to IM must be monitored.

5.3.1 Extraction and Purification

When analyzing IM in etiolated *Pinus sylvestris* seedlings, Sundberg et al. (1985) extracted the plant material with buffer rather than methanol in order to obtain cleaner samples. The plant material and appropriate internal standard (see Sect. 5.3.3) are homogenized in 0.1 M, pH 8.0 phosphate buffer (0.5 g ml^{-1}) containing 0.02% (w/v) sodium diethyldithiocarbamate as an antioxidant. The tissue homogenate is ultrasonicated for 15 min, filtered and saturated with ammonium sulphate to precipitate proteins. All steps are carried out at 4 °C. The extract is then subjected to ca. 45,000 *g* for 40 min in a refrigerated centrifuge, the supernatant is collected, adjusted to pH 8.0, slurried with PVP, filtered and run through an XAD-7 column which retains IM but not IAA. IM is eluted from the XAD-7 column with hexane/ethyl acetate (30:70, v/v). The aqueous phase of eluate is discarded and sodium diethyldithiocarbamate (0.02%) added to the organic phase prior to reduction to dryness in vacuo and analyses by HPLC (Fig. 9).

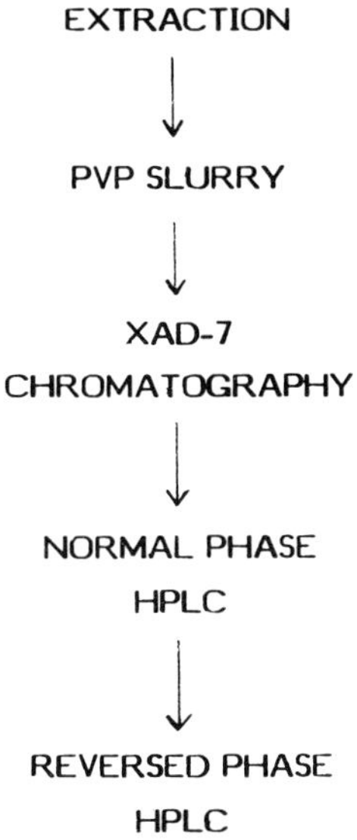

Fig. 9. Procedures for quantitative analysis of indole-3-methanol

5.3.2 HPLC Analysis

Sundberg et al. (1985) purified IM in *Pinus* extracts by normal phase HPLC on a 5 μm Nucleosil cyanopropyl support using a solvent of hexane/ethyl acetate (75:25, v/v) (Fig. 10). The putative IM was collected and reduced to dryness prior to analysis and quantification in a reversed phase system using a nonacidic buffer (Fig. 11). This comprised a 5 μm Nucleosil C-18 support and a mobile phase of 40% methanol in 10 mM, pH 6.5 phosphate buffer.

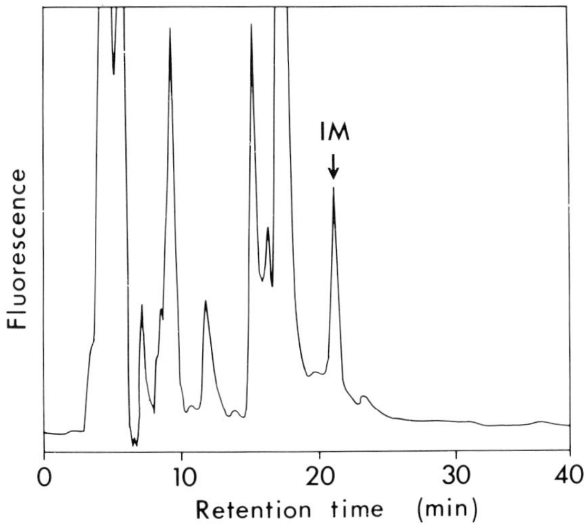

Fig. 10. Normal phase HPLC of a semi-purified extract from etiolated *Pinus sylvestris* seedlings. *Arrow* indicates peak co-chromatographing with indole-3-methanol. Column: 250 × 4 mm i.d. 5 μm Nucleosil cyanopropyl. Mobile phase: hexane/ethyl acetate (75:25, v/v). Flow rate: 0.9 ml min^{-1}

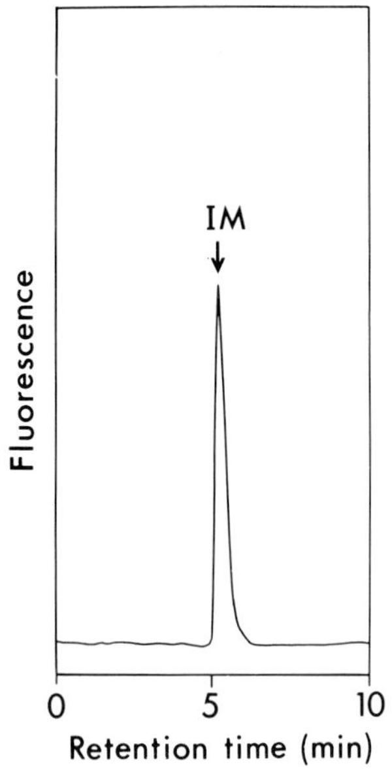

Fig. 11. Reversed phase ion-suppression HPLC of indole-3-methanol-like peak from Fig. 10. Column: 250 × 4 mm i.d. 5 μm Nucleosil C-18. Mobile phase: 40% methanol in 10 mM, pH 6.5 phosphase buffer. Flow rate: 0.6 ml min^{-1}

5.3.3 Internal Standards

In order to measure the conversion of IAA to IM during sample preparation, Sundberg et al. (1985) added 16.8×10^3 Bq [2-^{14}C]IAA (1.4 μg) to samples of etiolated *Pinus* seedlings at the initial extraction stage. In order to minimize the loss of any [^{14}C]IM formed, 1 μg IM was also added to the samples as a cold carrier. The extracts were processed as described above and both the IAA and IM fractions were analyzed by reversed phase HPLC. The relevant fluorescent peaks were collected and radioactivity measured by liquid scintillation counting. Assuming the same recovery of IM as IAA, it was calculated that the conversion of IAA to IM was only ca. 0.5%.

Quantification of IM in Scots pine was carried out by extracting 50 g of etiolated seedlings with 1.48×10^3 Bq [^{14}C]IM (107 ng) which was synthesized as described in Sect. 2. Extracts were purified and analyzed as outlined above in Sects. 5.3.1 and 5.3.2 and quantitative estimates of IM (V) were calculated to be 19.7 ± 1.4 ng g^{-1} ($\pm$SD) according to Eq. (4).

$$V = \frac{(\text{IM detected})\,([^{14}\text{C}]\text{IM added})}{([^{14}\text{C}]\text{IM detected})} - [(\text{sp. act.}\,[^{14}\text{C}]\text{IM})\,([^{14}\text{C}]\text{IM added})]\,. \tag{4}$$

The etiolated *Pinus* seedlings contained 175 ± 13.5 ng IAA g^{-1}, thus with an estimated 0.5% conversion of IAA to IM, it can be calculated that only 0.9 ng g^{-1} of the IM measured originated from IAA during sample preparation.

6 Conclusions

HPLC is a very flexible and practical tool for the quantitative analysis of IAA, ICA, and IM in *Pinus sylvestris* tissue. Accurate analysis is, however, very much dependent upon adequate sample purification. Whereever possible, radiolabeled internal standards should be used to monitor losses encountered during purification. When analyzing ICA and IM it is important to ascertain how much of the compound that is measured originates from IAA breakdown during sample preparation.

References

Allen JRF, Rivier L, Pilet P-E (1982) Quantification of indol-3-yl acetic acid in pea and maize seedlings by gas chromatography-mass spectrometry. Phytochemistry (Oxf) 21:525–530

Andersson B, Andersson K (1982) Use of Amberlite XAD-7 as a concentrator column in the analysis of endogenous plant growth hormones. J Chromatogr 242:353–358

Atsumi S, Kuraishi S, Hayashi T (1976) An improvement of auxin extraction procedure and its application to cultured plant cells. Planta (Berl) 129:245–247

Bearder JR (1980) Plant hormones and other growth substances – their background, structures and occurrence. In: Macmillan J (ed) Hormonal regulation of development. I. Molecular aspects of plant hormones. Encycl Plant Physiol New Ser, vol 9. Springer, Berlin Heidelberg New York, pp 9–112

Bandurski RS, Nonhebel HM (1984) Auxins. In: Wilkins MB (ed) Advanced plant physiology. Pitman, London, pp 1–20

Blakesley D, Hall JF, Weston GD, Elliott MC (1983) Simultaneous analysis of indole-3-acetic acid and detection of 4-chloroindole-3-acetic acid and 5-hydroxyindole-3-acetic acid in plant tissues by high-performance liquid chromatography of their 2-methylindolo-α-pyrone derivatives. J Chromatogr 258:155–164

Bowen DH, MacMillan J, Graebe JE (1972) Determination of specific radioactivity of [^{14}C]-compounds by mass spectroscopy. Phytochemistry (Oxf) 11:2253–2257

Burnett D, Audus LJ (1964) The use of fluorimetry in the estimation of naturally-occurring indoles in plants. Phytochemistry (Oxf) 3:396–415

Caruso JL, Zeisler CS (1983) Indole-3-acetic acid in Douglas fir seedlings: a reappraisal. Phytochemistry (Oxf) 22:589–590

Caruso JL, Smith RG, Smith LM, Cheng T, Daves GD (1978) Determination of indole-3-acetic acid in Douglas fir using a deuterated analog and selected ion monitoring. Comparison of microquantities in seedling and adult tree. Plant Physiol (Bethesda) 62:841–845

Cohen JD, Schulze A (1981) Double-standard isotope dilution assay. I. Quantitative assay of indole-3-acetic acid. Anal Biochem 112:249–257

Crozier A, Loferski K, Zaerr JB, Morris RO (1980) Analysis of picogram quantities of indole-3-acetic acid by high performance liquid chromatography-fluorescence procedures. Planta (Berl) 150:366–370

De Yoe DR, Zaerr JB (1976) Indole-3-acetic acid in Douglas fir. Analysis by gas-liquid chromatography and mass spectrometry. Plant Physiol (Bethesda) 58:299–303

Ehmann A (1977) The Van Urk-Salkowski reagent – a sensitive and specific chromatogenic reagent for silica gel thin-layer chromatographic detection and identification of indole derivatives. J Chromatogr 132:267–276

Gmelin R (1964) Occurence, isolation and properties of glucobrassicin and neoglucobrassicin. Colloq Int CNRS 123:159–168

Gmelin R, Virtanen AI (1961) Glucobrassicin, der Precursor von SCN^-, 3-Indolylacetonitril und Ascorbigen in *Brassica oleracea* species. Ann Acad Sci Fenn Ser II Chem 107:1–25

Haagen-Smit AJ, Dandliker WB, Wittwer SH, Murneek AE (1946) Isolation of 3-indoleacetic acid from immature corn kernels. Am J Bot 33:118–120

Hemberg T, Tillberg E (1980) The influence of extraction procedure on yield of indole-3-acetic acid in plant extracts. Physiol Plant 50:176–180

Iino M, Carr DJ (1982) Estimation of free, conjugated and diffusible indole-3-acetic acid in etiolated maize shoots by the indolo-α-pyrone fluorescence method. Plant Physiol (Bethesda) 69:950–956

Jensen E (1982) Analysis of indole derivatives by reversed-phase high-performance liquid chromatography. J Chromatogr 246:126–132

Letham DS, Goodwin PB, Higgins TV (1978) Phytohormones and related compounds: a comprehensive treatise. Elsevier/North-Holland Biomedical, Amsterdam

Libbert E, Kaiser W, Kunert R (1969) Interaction between plants and epiphytic bacteria regarding their auxin metabolism. VI. The influence of the epiphytic bacteria on the content of extractable auxin in the plant. Physiol Plant 22:432–439

Martin GC, Nishijima C, Labavitch JM (1980) Analysis of indoleacetic acid by the nitrogen-phosphorus detector gas chromatograph. J Am Soc Hortic Sci 105:46–50

Melhado LL, Pearce CJ, d'Alarcao M, Leonard NJ (1982) Specifically deuterated and tritiated auxins. Phytochemistry (Oxf) 21:2879–2885

Pengelly W, Meins F (1977) A specific radioimmunoassay for nanogram quantities of the auxin, indole-3-acetic acid. Planta (Berl) 136:173–180

Reeve DR, Crozier A (1977) Radioactivity monitor for high-performance liquid chromatography. J Chromatogr 137:271–282

Reeve DR, Crozier A (1980) Quantitative analysis of plant hormones. In: MacMillan J (ed) Hormonal regulation of development I. Molecular aspects of plants hormones. Encycl Plant Physiol New Ser, vol 9. Springer, Berlin Heidelberg New York, pp 203–280

Reeve DR, Crozier A (1983) HPLC radioactivity monitors – fact and fiction. Lab Pract 32(3):59–60

Rivier L, Pilet P-E (1974) Indolyl-3-acetic acid in cap and apex of maize roots: identification and quantification by mass fragmentography. Planta (Berl) 120:107–112

Sandberg G, Andersson B, Dunberg A (1981 a) Identification of 3-indoleacetic acid in *Pinus sylvestris* L. by gas chromatography-mass spectrometry, and quantitative analysis by ion-pair reversed-phase liquid chromatography with spectrofluorimetric detection. J Chromatogr 205:125–137

Sandberg G, Dunberg A, Odén P-C (1981 b) Chromatography of acid phytohormones on columns of Sephadex LH-20 and insoluble poly-N-vinylpyrrolidone, and application to the analysis of conifer extracts. Physiol Plant 53:219–224

Sandberg G, Jensen E, Crozier A (1984) Analysis of 3-indole carboxylic acid in *Pinus sylvestris* needles. Phytochemistry (Oxf) 23:99–102

Sandberg G, Ljung K, Alm P (1985) Precision and accuracy of radioimmunoassay in the analysis of 3-indoleacetic acid in needles of Scots pine. Phytochemistry (Oxf) 24:1439–1442

Seeley SD, Powell LE (1974) Gas chromatography and detection of microquantities of gibberellins and indoleacetic acid as their fluorinated derivatives. Anal Biochem 58:39–46

Sembdner G, Gross D, Liebisch H-W, Schneider G (1980) Biosynthesis and metabolism of plant hormones. In: MacMillan J (ed) Hormonal regulation of development I. Molecular aspects of plant hormones. Encycl Plant Physiol New Ser, vol 9. Springer, Berlin Heidelberg New York, pp 281–444

Srivastava BIS (1963) Ether-soluble and ether-insoluble auxins from immature corn kernels. Plant Physiol (Bethesda) 38:473–478

Stoessl A, Venis MA (1970) Determination of submicrogram levels of indole-3-acetic acid. A new, highly specific method. Anal Biochem 34:344–351

Sundberg B, Sandberg G, Jensen E (1985) Identification and quantification of indole-3-methanol in Scots pine (*Pinus sylvestris* L.). Plant Physiol (Bethesda) 77:952–955

Sundberg B, Sandberg G, Crozier A (1986) Purification of indole-3-acetic acid in plant-extracts by immunoaffinity chromatography. Phytochemistry (Oxf) 25:295–298

Sweetser PB, Swartzfager DG (1978) Indole-3-acetic acid levels in plant tissue as determined by a new high performance liquid chromatographic method. Plant Physiol (Bethesda) 61:254–258

Weiler EW (1981) Radioimmunoassay for picomol quantities of indole-3-acetic acid for use with highly stable ^{125}I and ^{3}H-IAA derivatives as radiotracers. Planta (Berl) 153:319–325

Weiler EW (1982) Plant hormone immunoassay. Physiol Plant 54:230–234

Weiler EW, Jourdan PS, Conrad W (1981) Levels of indole-3-acetic acid in intact and decapitated coleoptiles as determined by a specific and highly sensitive solid-phase enzyme immunoassay. Planta (Berl) 153:561–571

Wurst M, Prikryl Z, Vancura V (1980) High-performance liquid chromatography of plant hormones. I. Separation of plant hormones of the indole type. J Chromatogr 191:129–136

Wurst M, Prikryl Z, Vokoun J (1984) High-performance liquid chromatography of plant hormones. II. Determination of plant hormones of the indole type. J Chromatogr 286:237–245

HPLC for the Separation and Determination of Phenolic Compounds in Plant Cell Walls

R. D. HARTLEY

1 Introduction

There is growing interest in the chemical composition of plant cell walls resulting from emphasis being placed on the utilisation of biomass and on the use of "dietary fibre" in human nutrition. In many countries cereal crop residues are waste products, and some of these may be utilised after treatment to increase the degradability of their cell walls, which usually account for a large proportion of the dry weight (for example approximately 80% of the dry matter of barley or wheat straw). The biodegradability of cell walls, particularly graminaceous cell walls, is often limited by phenolic constituents (Hartley 1981, 1985).

The chemical nature of lignin in non-woody plant materials is poorly characterised, but in recent years it has been established that several monocotyledonous families including the Gramineae (grasses and cereals) have phenolic acids and aldehydes linked to their cell walls (see Fig. 1) (e.g. El-basyonni et al. 1964; Fulcher et al. 1972; Hartley 1972, 1973; Fincher 1976; Hartley and Jones 1976; Markwalder and Neukom 1976; Harris and Hartley 1976, 1980; Harris et al. 1980; Akin et al., in press). Graminaceous cell walls contain 1–3% of their dry matter in the form of these phenolic compounds, mainly trans-ferulic and trans-p-coumaric acids.

It has recently been shown that in sugar cane and maize cell walls, ferulic acid is ester-linked to arabinose, which in turn is linked to xylose (Kato et al. 1983; Kato and Nevins 1985). Very recently Mueller-Harvey et al. (1986) have established that the binding of ferulic and p-coumaric acids to cell walls of barley straw are as shown in Fig. 2 (P = p-coumaroyl, F = feruloyl, A = arabinosyl, X′, X = xylosyl): the linkages are the same as those shown earlier for ferulic acid in the cell walls of sugar cane and maize. The two compounds shown in Fig. 2 were isolated from enzymic hydrolysates of cell walls.

In general, cell walls of dicotyledons have only very small amounts of these phenolic acids associated with their cell walls, but species of the Caryophyllales, which include the sugar and fodder beets, contain amounts which approach those of the monocotyledons (Hartley and Harris 1981).

This chapter describes HPLC techniques for the determination and preparation of the phenolic acids and the three aldehydes of cell walls (Fig. 1). Similar techniques may be used for the determination of the aldehydes produced when cell walls or lignin preparations are oxidised with alkaline nitrobenzene or alkaline cupric oxide. These oxidation methods have often been used to examine plant lignins (e.g. Brand 1967; Higuchi et al. 1967; Pepper et al. 1967; Hartley 1971).

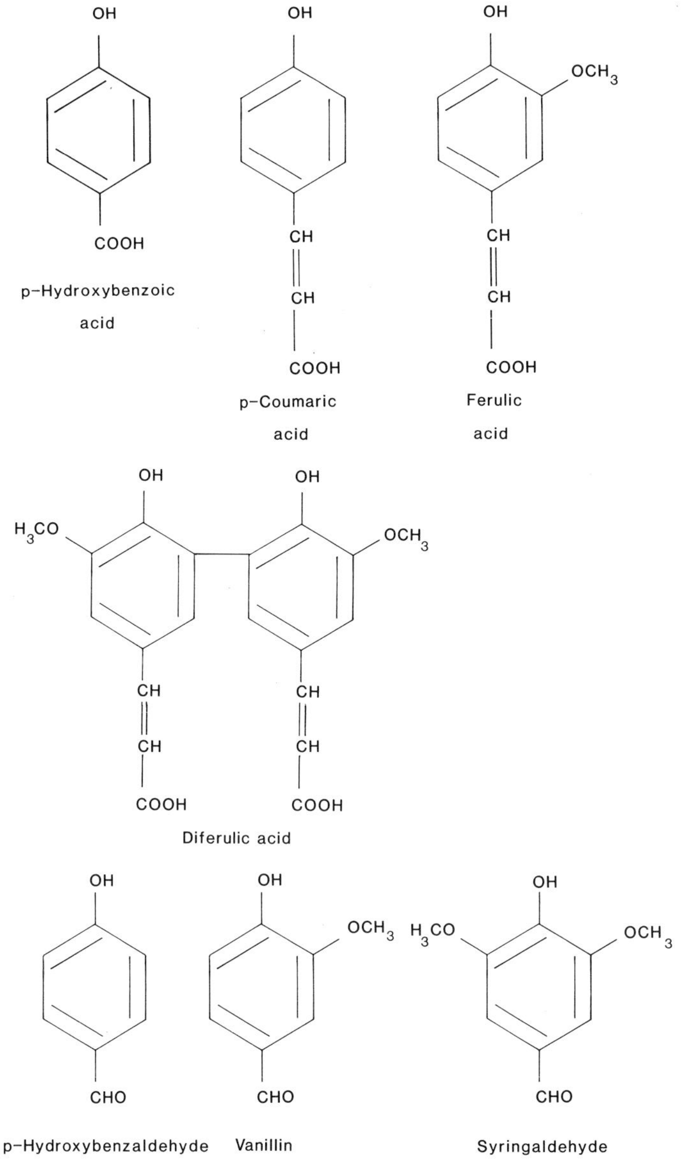

Fig. 1. Structural formulae of phenolic acids and aldehydes bound to cell walls

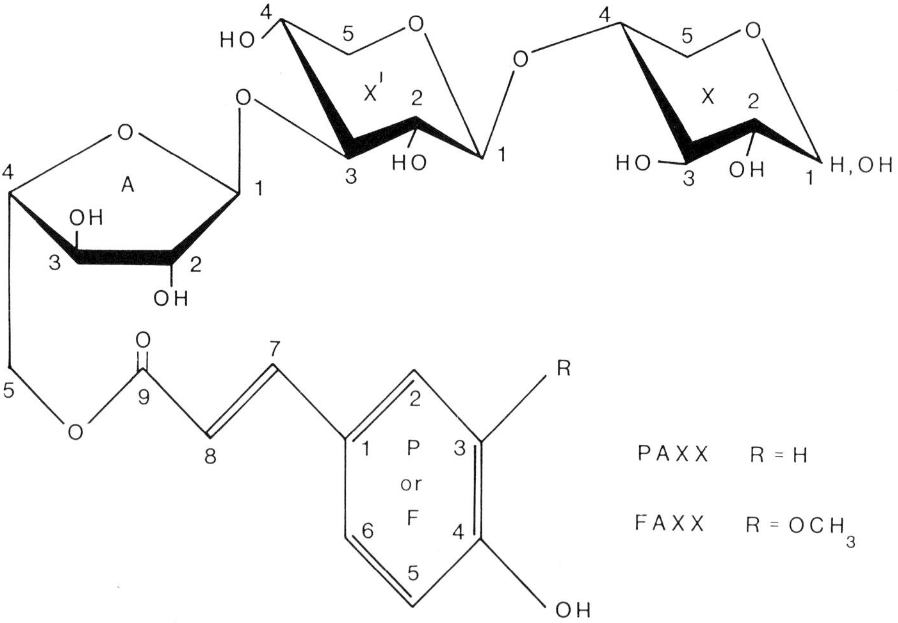

Fig. 2. Binding of p-coumaric and ferulic acids to sugar residues of graminaceous cell walls

2 General Considerations

The use of high performance liquid chromatography (HPLC) has several advantages over other methods for determination of aromatic compounds in cell walls. The technique is sensitive, rapid and does not require the preparation of derivatives before separation. A further advantage is that ultraviolet absorption detectors set in the aromatic region of the spectrum (e.g. 275 nm) may be used. Thus compounds that do not adsorb in this region are not recorded, allowing "impure" mixtures (e.g. those containing sugars) to be successfully analysed. Several reports have been published of the application of HPLC to various phenolic acids and aldehydes that occur in plants (e.g. Wulf and Nagel 1976; Murphy and Stutte 1978; Hartley and Buchan 1979; Roston and Kissinger 1982; Graham and Aman 1984; Akin et al., in press). Gas-liquid chromatography (GLC) of phenols and acids has been used by several groups but requires the preparation of derivatives and is slower (e.g. Hartley and Jones 1975; Salomonsson et al. 1978; Van Sumere et al. 1979).

In order to isolate cell walls from many plant materials, a procedure is recommended that is based on a modification (Hartley et al. 1974) of the neutral detergent method of Van Soest and Wine (1967). A major attraction of this technique is the speed of cell wall preparation, one operator being able to isolate as many as 50 wall preparations (0.5 g) from dried, ground plant materials in 1 day. An alternative but slower technique is that of Harris and Hartley (1980). In some circumstances it might be preferred because, unlike the detergent method, no heat is required and at least 95% of the cells are ruptured to release soluble non-wall components. In our laboratories the two methods have given similar results for the determination of cell wall phenolic compounds.

Several other techniques have been described for the isolation of plant cell walls (dietary fibre) from foods (e.g. Southgate 1969; Robertson and Van Soest 1981; Asp et al. 1983; Englyst and Cummings 1984): these involve treatment with α-amylase in order to remove the large amounts of starch that occur in many foods. Such techniques of dietary fibre preparation are suitable for the determination of wall phenolic acids and aldehydes, provided these compounds are not released or degraded by the reagents employed. The technique described below for the isolation of dietary fibre is based on the recommended procedure for plant cell walls, but incorporates added treatments with α-amylase to remove starch and protease to remove protein (see also Robertson and Van Soest 1981).

The procedure recommended for the release by alkali treatment of phenolic acids and aldehydes from cell walls and their determination by HPLC (Sects. 4 and 5) are based on earlier work (Hartley and Buchan 1979). It is important that all manipulations of solutions are undertaken in the absence of ultraviolet radiation in order to prevent cis-trans isomerisation (Neish 1961; Kahnt 1967; Hartley and Jones 1975): "white" fluorescent tubes are suitable (Hartley and Buchan 1979).

3 The Recommended Procedure – Isolation of Cell Walls

3.1 Reagents and Equipment

3.1.1 Reagents (All Analytical Grade)

1. Neutral detergent solution (1 l)
   Dissolve sodium dodecyl sulphate (30.0 g) plus ethylene diamine tetracetic acid, disodium salt (18.0 g) plus disodium tetraborate decahydrate (6.81 g) plus 2-ethoxyethanol (10.0 ml) in glass distilled water by heating and dilute with water to 1 l. Filter (glass sinter porosity 1, 100–120 μm) Check pH is 7.0 ± 0.1 (adjust with NaOH or H_2SO_4 if necessary).
2. Distilled water
3. Acetone
4. Ether

3.1.2 Equipment

1. Freeze-dryer
2. Hammer mill equipped with a 0.8 mm mesh screen
   (e.g. Micro Hammer Mill, Glencreston, London)
3. Hot plate
4. Conical flasks (500 ml)

5. "Cold finger" reflux system to fit each conical flask
   to enable aqueous refluxing to be undertaken
   (e.g. Gallenkamp, London)
6. Glass sintered crucibles (capacity 30 ml, porosity 1, 100–120 μm)
7. Buchner flasks (1 l)
8. Source of suction (e.g. water pump)
9. Desiccator (silica gel)

3.2 Isolation of Cell Walls

The plant material is freeze-dried and ground to pass the 0.8-mm mesh screen. A sample (1.0 g) is shaken with neutral detergent solution (100 ml) in a conical flask and refluxed for 1 h with occasional shaking to prevent plant material from sticking to the bottom or side of the flask. The suspension is filtered by transferring to a tared glass sintered crucible fitted to a Buchner flask. The cell walls are washed thoroughly with hot water (ca. 80 °C, 60 ml) followed by acetone (60 ml) then ether (60 ml), before air-drying (to remove ether). The walls are then dried over silica gel to attain constant weight.

3.3 Isolation of Dietary Fibre

The freeze-dried plant material (1.0 g) is ground and extracted with neutral detergent solution as described in Sect. 3.2. Starch is removed from the crude cell walls (1.0 g from several extractions) by shaking with α-amylase solution (20 ml) at 37 °C using filter tubes fitted with rubber teats as described in Sect. 4.2 (or in conical flasks): the enzyme solution is prepared by stirring α-amylase (Sigma Type IIA, 20 mg) with water (100 ml). A small volume of toluene is added to each suspension to prevent fungal and bacterial growth. Removal of starch is monitored by microscopy (iodine stain): for example, walls of wheat bran, wheat and oat grain were incubated for 48 h and those of maize grain for 96 h (Hartley et al., unpublished). The suspensions are filtered as described in Sect. 4.2 (or glass sintered crucibles, porosity 1, are used if conical flasks are employed for incubation), washed with water (60 ml) and excess water removed by applying pressure above the filter. The cell walls are washed with acetone (60 ml) then ether (60 ml), and air-dried before final drying over silica gel.

If required, protein may be removed from α-amylase-treated cell walls (300 mg) by shaking with protease solution (20 ml) plus a small volume of toluene at 37 °C for at least 48 h: the enzyme solution is prepared by shaking protease (Sigma Type XIV, 2.5 g) with sodium phosphate buffer (1 l 0.01 M containing 0.10 g NaN_3) at 20 °C for 15 min. The suspensions are filtered and the residue washed with water, acetone and ether, then dried as described above (Sect. 3.2).

4 The Recommended Procedure – Release of Phenolic Acids and Aldehydes from Cell Walls and Their Separation and Determination by HPLC (C_{18} Reverse Phase) *(Methods 1 and 2)*

4.1 Reagents and Equipment

4.1.1 Reagents (All Analytical Grade)

1. Distilled water.
2. 1 M NaOH.
3. 6 M HCl.
4. Nitrogen ("oxygen-free").
5. p-Anisic acid (100 mg) dissolved in methanol : water (450 ml : 50 ml).

4.1.2 Equipment

1. Glass filter tubes (capacity ca. 3 ml, ca. 1 cm internal diameter) fitted with a sintered disc, porosity 3, 20–30 μm, female ground glass joint (B.10) and stopper (B.10).
2. Rubber teats to seal the ends of the filter tubes.
3. Incubator (set at 20 °C).
4. Shaker to rotate flasks end-over-end (e.g. Labquake, Labindustries, Berkeley, California)
5. Blow-ball fitted to a male joint (B.10) to enable pressure to be applied to suspensions in the filter tubes to decrease the time of filtration.
6. Volumetric flasks (5 ml).

4.2 Release of Phenolic Acids and Aldehydes from Cell Walls and Preparation of Solutions for Analytical HPLC (C_{18} Reverse Phase)

Cell walls (30 mg) are weighed into a filter tube fitted with a rubber teat at its lower end. 1 M NaOH (2.0 ml) is added, air removed from the tube by flushing with nitrogen and the stopper secured. The suspension is shaken at 20 °C for 20 h. The rubber teat is carefully removed, the filter tube placed in the neck of the volumetric flask and the suspension filtered (if necessary filtration can be speeded up by applying pressure with the blowball). The residue is washed with water (total volume of filtrate ca. 2 ml). The filtrate is acidified to pH 2.5 with 6 M HCl. p-Anisic acid solution (500 μl) is added as internal reference and the filtrate diluted to 5.0 ml with water.

4.3 Analytical HPLC (C_{18} Reverse Phase) of Phenolic Acids and Aldehydes *(Methods 1 and 2)*

Water to be used in the preparation of the HPLC solvent mixture is glass distilled. Acetic acid and n-butanol are of analytical grade. The solvent mixture is filtered (ca. 0.45 µm membrane filter) and may be conveniently degassed by bubbling through helium for approximately 15 min.

HPLC is carried out using a high pressure pump (e.g. Constametric IIIG, Laboratory Data Control, Riviera Beach, Florida), a loop injector (e.g. Rheodyne 7125, Cotati, California), a variable wavelength detector (e.g. Spectromonitor III, LDC) set at 275 nm with 10 µl flow cell, and an integrator and printer-plotter (e.g. CI-10B, LDC). An aliquot (ca. 50 µl) of the filtrate is injected on to a steel column (25×0.46 cm internal diameter) containing Spherisorb C_{18} bonded-phase on silica (Phase Separations, Queensferry, UK; ODS-1, 5 µm) followed by isocratic elution with degassed water-glacial acetic acid-n-butanol (342:1:14, v/v, *method 1*). The flow rate is 1.2 ml min^{-1} and the column pressure ca. 1.3×10^7 Pa.

Reference mixtures of the acids (excluding cis-ferulic acid which is not available commercially) are chromatographed by the same method and shown to give linear responses to the detector in the working range (0–1.2 µg of each compound). cis-Ferulic acid is estimated assuming the same response to the detector as that of the trans isomer.

The separation of both phenolic acids and aldehydes can be achieved by the same method except that the eluting solvent is water-glacial acetic acid-n-butanol (347:1:11, v/v), this being designated *method 2*. Method 1 is preferred for the separation of p-coumaric and ferulic acids which are the major phenolic constitutents released from plant cell walls. Method 2 is mainly employed for the examination of the phenolic components of soils (Hartley and Buchan 1979; Whitehead et al. 1981, 1982).

Retention times of phenolic acids and aldehydes using methods 1 and 2 are shown in Table 1.

5 Alternative Procedure – Release of Phenolic Acids and Aldehydes from Cell Walls and Their Separation and Determination by HPLC (PRP-1 Resin) *(Methods 3 and 4)*

5.1 Reagents and Equipment

As described in Sect. 4.1 plus methylene chloride and anhydrous sodium sulphate (both analytical grade) and filter papers (Whatman No. 1).

5.2 Release of Phenolic Acids and Aldehydes from Cell Walls and Preparation of Solutions for HPLC (PRP-1 Resin)

5.2.1 Method for Analytical HPLC (PRP-1 Resin)

This is described in Sect. 4.2. The filtrate is adjusted with HCl to the pH of the solvent used.

5.2.2 Method for Preparative HPLC (PRP-1 Resin)

All solvents are analytical grade.

The method is similar to that described under Sect. 4.2 except that cell walls (100 mg) are treated with 1 M NaOH (5 ml) in larger capacity filter tubes (ca. 15 ml). A minimum of water is used to wash the residue from filtration. The filtrate is acidified with 6 M HCl to pH 2.5 and extracted with methylene chloride (3 × 10 ml). The extract is dried by the addition of anhydrous sodium sulphate and filtered (Whatman No. 1). The solvent is removed by evaporation in a stream of "oxygen-free" nitrogen and the residue dissolved in 0.01 M NaOH (1 ml) containing 0.4 M NaCl.

5.3 Analytical HPLC (PRP-1 Resin) of the Phenolic Acids and Aldehydes

5.3.1 Phenolic Aldehydes and trans, trans-Diferulic Acid *(Method 3)*

The method is similar to that described in Sect. 4.3 except that the column (25 × 0.46 cm internal diameter) is packed with PRP-1 resin (10 μm, Hamilton, Reno, Nevada), the solvent is 0.01 M NaOH containing 0.4 M NaCl, the flow rate 0.8 ml min^{-1}, ca. 3.5×10^6 Pa pressure, and the ultraviolet absorption detector, set at 275 nm.

Retention times are shown in Table 1.

5.3.2 p-Coumaric and Ferulic Acids *(Method 4)*

The method is similar to that described above except that 0.1 M $NaHCO_3$-methanol (19:1, v/v) solvent (pH 8.6) is employed. Retention times are shown in Table 1.

Table 1. Retention times (min) of phenolic acids and aldehydes released from plant cell walls by treatment with NaOH

	HPLC method			
	1	2	3	4
p-Hydroxybenzoic acid		9.8		
Cis-p-Coumaric acid	13.2	16.0		9.5
Trans-p-Coumaric acid	21.0	24.5		10.5
Cis-Ferulic acid	17.6	21.0		14.8
Trans-Ferulic acid	24.8	31.7		22.6
Trans,trans-Diferulic acid			14.0	
p-Hydroxybenzaldehyde		11.1	11.7	
Vanillin		16.5	21.2	
Syringaldehyde			46.0	

Table 2. Amounts of phenolic acids (mg g^{-1}) released from dry plant cell walls by treatment with NaOH (HPLC method 1)

	p-Coumaric acid		Ferulic acid		Ref.
	trans	cis	trans	cis	
Italian ryegrass, shoot	0.61	0.27	5.62	1.20	Hartley and Vaverkamp (1984
Barley straw, whole sample	2.74	0.26	3.72	0.00	
Barley straw, stem	3.02	0.17	3.42	0.09	
Maize (normal)					
leaf blade	4.42	1.00	3.50	0.36	
leaf sheath	9.32	1.06	6.49	0.62	
stem	24.25	2.54	5.71	0.56	
Maize (mutant bm_3)					
leaf blade	2.01	0.69	3.54	0.56	
leaf sheath	4.70	0.72	6.26	0.76	
stem	12.99	1.30	5.77	0.59	
Wheat bran	0.15	0.03	4.81	1.80	
Perennial ryegrass, shoot[a]	0.88	0.66	4.54	1.49	Hartley and Dibb (unpublished)[b]
Tall fescue hay[a]	3.73	1.20	4.26	0.59	
Lucerne, shoot	0.02	0.03	0.02	0.00	
Oat grain	1.99	0.33	4.71	0.41	
Maize grain	2.57	0.46	18.03	2.63	
Pineapple	2.09	0.51	17.68	4.18	

[a] HPLC method 2 showed perennial ryegrass and tall fescue to contain 0.21 and 0.06 mg p-hydroxybenzoic acid g^{-1} cell walls respectively.
[b] Cell walls isolated as described by Harris and Hartley (1980) and treated with α-amylase followed by protease (see Sect. 3.3).

Table 3. Amounts of phenolic acids and aldehydes (mg g^{-1}) released from dry plant cell walls by treatment with NaOH (HPLC methods 3 and 4)[a]

	Maize stem	Barley straw stem	Wheat bran
Trans+cis-p-coumaric acid[b]	33.05	3.42	0.07
Trans-ferulic acid	3.78	2.26	3.27
Cis-ferulic acid	0.00	0.00	2.33
Trans,trans-diferulic acid	0.11	0.13	0.18
p-Hydroxybenzaldehyde	0.26	0.02	0.06
Vanillin	0.22	0.23	0.10
Syringaldehyde	0.22	0.10	0.00

[a] Hartley and Keene (1984).
[b] Incomplete resolution of trans and cis isomers.

5.4 Preparative HPLC (PRP-1 Resin) of the Phenolic Acids and Aldehydes

5.4.1 Preparation of the Aldehydes and trans,trans-Diferulic Acid

The method of separation is as described in Sect. 5.3 except that ca. 100 μl of the solution containing the residue is injected on to the column and the detector is "off-set" at 340 nm to decrease its sensitivity and so enable the separation to be monitored. Collected elutates containing the separated compounds are acidified to pH 2.5 with 6 M HCl and the solution extracted with methylene chloride (3 × 1 vol). The organic layer is dried with anhydrous sodium sulphate, filtered (Whatmann No. 1) and the solvent removed in a stream of "oxygen-free" nitrogen.

5.4.2 Preparation of p-Coumaric and Ferulic Acids

The method is similar to that described above except that $NaHCO_3$-methanol (19 : 1, v/v) solvent (see Sect. 5.3.2) is employed and the extract (see Sect. 5.2.2) is adjusted to pH 8.6 with 6 M HCl before injection on to the column. The acidified eluates containing the acids are extracted with ether instead of methylene chloride.

It should be noted that this method does not completely separate the cis and trans isomers of p-coumaric acid. If separation of these isomers is required, then the reverse phase HPLC method should be employed (see Sects. 4.2 and 4.3).

6 Amount of Phenolic Acids and Aldehydes Released from the Cell Walls of Graminaceous Plants

Results obtained using the methods described in this chapter are shown in Tables 2 and 3.

Acknowledgments. I wish to thank Mrs. P. L. Burrows for the preparation of the typescript and Dr. D. C. Whitehead for his helpful comments.

References

Akin DE, Hanna WW, Snook ME, Himmelsbach DS, Barton FE, Windham WR (1986) Chemical variation and digestibility in normal 12 and bmr 12 sorghum. Agron J (in press)

Asp NG, Johansson C-G, Hallmer H, Siljestrom M (1983) Rapid enzymatic assay of insoluble and soluble dietary fiber. J Agric Food Chem 31:476–482

Brand JM (1967) Studies of grass lignins. II. The estimation of lignin oxidation products by gas liquid chromatography. J Chromatogr 26:373–377

El-basyouni SZ, Neish AC, Towers GHN (1964) The phenolic acids in wheat. III. Insoluble derivatives of phenolic cinnamic acids as natural intermediates in lignin biosynthesis. Phytochemistry (Oxf) 3:627–639

Englyst HN, Cummings JH (1984) Simplified method for the measurement of total non-starch polysaccharides by gas-liquid chromatography of constituent sugars as alditol acetates. Analyst 109:937–942

Fincher GB (1976) Ferulic acid in barley cell walls; a fluorescence study. J Inst Brew 82:347–349

Fulcher RG, O'Brien TP, Lee JW (1972) Studies of the aleurone layer. I. Conventional and fluorescence microscopy of the cell wall with emphasis on phenol-carbohydrate complexes in wheat. Aust J Biol Sci 25:23–34

Graham H, Åman P (1984) A comparison between degradation in vitro and in sacco of constituents of untreated and ammonia-treated barley straw. Anim Feed Sci Technol 10:199–211

Harris PJ, Hartley RD (1976) Detection of bound ferulic acid in cell walls of the Gramineae by ultraviolet fluorescence microscopy. Nature 259:508–510

Harris PJ, Hartley RD (1980) Phenolic constituents of the cell walls of monocotyledons. Biochem Syst Ecol 8:153–160

Harris PJ, Hartley RD, Lowry KH (1980) Phenolic constituents of mesophyll and non-mesophyll cell walls from leaf laminae of *Lolium perenne*. J Sci Food Agric 31:959–962

Hartley RD (1971) Improved methods for the estimation by gas-liquid chromatography of lignin degradation products from plants. J Chromatogr 54:335–344

Hartley RD (1972) p-Coumaric and ferulic acid components of cell walls of ryegrass and their relationships with lignin and digestibility. J Sci Food Agric 23:1347–1354

Hartley RD (1973) Carbohydrate esters of ferulic acid as components of cell walls of *Lolium multiflorum*. Phytochemistry (Oxf) 12:661–665

Hartley RD (1981) Chemical constitution, properties and processing of lignocellulosic wastes in relation to nutritional quality for animals. Agric Environ 6:91–113

Hartley RD (1985) Chemistry of lignocellulosic plant materials and non-microbial processes for increasing their feed value for the ruminant. OECD workshop Improved utilisation of lignocellulosic materials with special reference to animal feed. OECD Paris, pp 10–30

Hartley RD, Buchan H (1979) High performance liquid chromatography of phenolic acids and aldehydes derived from plants or from the decomposition of organic matter in soil. J Chromatogr 180:139–143

Hartley RD, Harris PJ (1981) Phenolic constituents of the cell walls of dicotyledons. Biochem Syst Ecol 9:189–203

Hartley RD, Haverkamp J (1984) Pyrolysis-mass spectrometry of the phenolic constituents of plant cell walls. J Sci Food Agric 35:14–20

Hartley RD, Jones EC (1975) Effect of ultraviolet light on substituted cinnamic acids and the estimation of their cis and trans isomers by gas chromatography. J Chromatogr 107:213–218

Hartley RD, Jones EC (1976) Diferulic acid as a component of cell walls of *Lolium multiflorum*. Phytochemistry (Oxf) 15:1157–1160

Hartley RD, Keene AS (1984) Aromatic aldehyde constituents of graminaceous cell walls. Phytochemistry (Oxf) 23:1305–1307

Hartley RD, Jones EC, Fenlon JC (1974) Prediction of the digestibility of forages by treatment of their cell walls with cellulolytic enzymes. J Sci Food Agric 25:947–954

Higuchi T, Ito Y, Kawamura I (1967) p-Hydroxyphenylpropane component of grass lignin and role of tyrosine-ammonia lyase in its formation. Phytochemistry (Oxf) 6:875–881

Kahnt G (1967) Trans-cis-equilibrium of hydroxycinnamic acids during irradiation of aqueous solutions at different pH. Phytochemistry (Oxf) 6:755–758

Kato A, Azuma J, Koshijima T (1983) A new feruloylated trisaccharide from bagasse. Chem Lett (Chem Soc Jpn) 137–140

Kato Y, Nevins DJ (1985) Isolation and identification of O-(5-O-feruloyl-α-L-arabinofuranosyl)-(1→3) -O-β-D-xylopyranosyl-(1→4)-D-xylopyranose as a component of *Zea mays* shoot cell walls. Carbohydr Res 137:139–150

Markwalder HU, Neukom H (1976) Diferulic acid as a possible crosslink in hemicelluloses from wheat germ. Phytochemistry (Oxf) 15:836–837

Mueller-Harvey IM, Hartley RD, Harris PJ, Curzon EH (1986) Linkage of p-coumaryl and feruloyl groups to cell wall polysaccharides of barley straw. Carbohydr Res 148:71–85
Murphy JB, Stutte CA (1978) Analysis for substituted benzoic and cinnamic acids using high-pressure liquid chromatography. Anal Biochem 86:220–228
Neish AC (1961) Formation of m and p-coumaric acids by enzymatic deamination of the corresponding isomers of tyrosine. Phytochemistry (Oxf) 1:1–24
Pepper JM, Casselman BW, Karapolly JC (1967) Lignin oxidation. Preferential use of cupric oxide. Can J Chem 45:3009–3017
Robertson JB, Van Soest PJ (1981) The detergent system of analysis and its application to human foods. In: James WPT, Theander O (eds) The analysis of dietary fiber. Dekker, New York, pp 123–158
Roston DA, Kissinger PT (1982) Liquid chromatographic determination of phenolic acids of vegetable origin. J Liq Chrom 5 Suppl 1:75–103
Salomonsson A-C, Theander O, Åman P (1978) Quantitative determination by glc of phenolic acids as ethyl derivatives in cereal straws. J Agric Food Chem 26:830–835
Southgate DAT (1969) Determination of carbohydrates in foods. J Sci Food Agric 20:331–335
Van Soest PJ, Wine RH (1967) Use of detergents in the analysis of fibrous feeds. IV. Determination of plant cell-wall constituents. J Assoc Off Anal Chem 50:50–55
Van Sumere CF, Van Brussel W, Vante Casteele K, Van Rompaey L (1979) Recent advances in the separation of plant phenolics. In: Swain T, Harborne JB, Van Sumere CF (eds) Biochemistry of plant phenolics. Plenum, New York, pp 1–28
Whitehead DC, Dibb H, Hartley RD (1981) Extractant pH and the release of phenolic compounds from soils, plant roots and leaf litter. Soil Biol Biochem 13:343–348
Whitehead DC, Dibb H, Hartley RD (1982) Phenolic compounds in soil as influenced by the growth of different plant species. J Appl Ecol 19:579–588
Wulf LW, Nagel CW (1976) Analysis of phenolic acids and flavonoids by HPLC. J Chromatogr 116:271–279

HPLC of Thiophenes for Phytochemical and Biochemical Research

R. SÜTFELD

1 Introduction

Thiophene derivatives are known from either synthetic or natural sources. Their occurrence in the Asteraceae family and in certain fungi is well documented (Bohlmann et al. 1973). They may be roughly classified by subdivision into mono-, di-, and terthiophenes; thiophene ring systems of naturally occurring compounds are usually connected at the 2,2′-positions. By substitution with various functional groups to thiophenes, a high diversity of chemical characters can be obtained; the most striking feature about natural thiophenes is the substitution with acetylenic side chains. For these reasons, most different chemical characteristics regarding solubility (polarity), distinct spectral properties as well as a remarkable instability (see below) of most of these compounds may be explained.

For many years, thiophene derivatives have been extracted from plant material at room temperature using diethylether/petrolether (EtOEt/PE) mixtures, followed by SiO_2 column- and/or thin-layer chromatography. Identification of purified compounds was performed by UV-, IR-, MS-, and NMR spectroscopy. The analytical data obtained in this way were fixed in numerous publications which were recently reviewed by Bohlmann et al. (1973). Also, another review from Sørensen (1977) dealt with the occurrence as well as with chemical characters of thiophene derivatives in plant systems.

One may criticize the fact that in these former investigations the instability of most of the naturally occurring thiophenes was either not or only minimally considered during the extraction/purification procedures employed. For example, the decomposition of photoreactive compounds like dithienylbutinene **11** or terthiophene **13** (see Fig. 3) is dependent on the time of exposure to light and temperature. Polar solvents also affect the stability of thiophenes to a great extent (Sütfeld 1984). Particularly under illumination with long-wave UV light (UV-A), such compounds tend to interact with others in biological systems (Towers 1980; Arnason et al. 1983) or polymerize after short exposure to UV-A in different solvents (Sütfeld 1984).

Therefore, new ways of extraction and purification procedures have to be considered to avoid failures and irreproducibilities, for example, in quantitative chemotaxonomical or in enzymological studies. Several possible solutions of this problem present themselves: (a) deep-temperature extraction under exclusion of day- and of UV light; (b) use of HPLC as a fast and careful purification technique; (c) storage and handling of extracts and of purified material under appropriate conditions. The following report intends to answer questions regarding some of these aspects.

2 Material and Methods

All experiments described here were performed by use of a ternary Spectra-Physics (Santa Clara, CA, USA) SP 8700 HPLC system. Detection was spectrophotometrical, using either a Gilford 250 spectrophotometer, equipped with an HELLMA (Müllheim, FRG) universal flow-thru cell (10 mm pathlength), or a KNAUER (Bad Homburg, FRG) Nr. 731.87 Digital-Spectrophotometer (10 mm pathlength) or a KRATOS (Ramsey, NJ, USA) Spectroflow 757 Absorbance Detector (8 mm or 1 mm pathlength cell, respectively). All these systems proved suitable for analytical as well as for semi-preparative HPLC of thiophene derivatives. Peak integration was performed by use of a Spectra-Physics SP 4100 computing integrator. Chromatograms were recorded by a Philips N2541 cassette recorder; this system allows storage of data and later playback of chromatograms with variable amplification or attenuation for comparison purposes.

A semi-preparative column (8 × 250 mm), filled with LiChrosorb (MERCK, Darmstadt, FRG) RP18, 5 μ, was purchased from KNAUER. Analytical columns (4.0 × 250 mm and 4.6 × 250 mm) were self-filled using the SP 8700 and a KNAUER filling column: The amount of filling material in question [e.g., LiChrosorb RP18, 5 μ, or Spherisorb (Queensferry, UK) ODS II, 5 μ) was slurried in degassed CH_3CN and was then degassed under vacuum. The column equipment was filled with the slurry at the highest possible pressure (about 39 MPa). Columns prepared in this manner showed about the same fractionation qualities as commercially available ones. For most purposes, guard columns (4 × 30 mm), dry-filled with LiChrosorb RP8, 10 μ, were used to avoid contamination of the column with insoluble material.

All chemicals used were of the highest grade of purity in order to avoid background noise or background shifts on gradient runs: EtOH (for extraction), UV-grade (BAKER, Deventer, The Netherlands); CH_3CN, HPLC grade (ROTH, Karlsruhe, FRG); H_2O, HPLC grade (BAKER). CH_3CN/H_2O residues from HPLC-runs were collected and bi-distilled over P_2O_5. This distillate proved to be an excellent solvent for HPLC purposes. It contains about 95% CH_3CN and shows optical qualities the same as or better than commercially available CH_3CN. In the following, this distillate is called CH_3CN–H_2O-Azeotrop.

All extraction procedures were performed under exclusion of UV or daylight. Some aspects of fundamental practice have already been described (Sütfeld 1982; Sütfeld and Towers 1982; Pensl and Sütfeld 1985). Plant material from various sources was deep-frozen in the respective homogenization vessel (mortar or blendor chamber) at −70 °C. Homogenization was performed using ethanol (EtOH) (−40 °C) in a mortar (with addition of quartzsand) or in a Waring blendor. The extract was allowed to warm up until −10 °C and insoluble residues were filtered off; H_2O was added in a ratio of 1 : 1 and products were extracted into petrolether (PE) (M.P. 40°–60 °C) immediately. After evaporation (condenser temp. −40 °C; sample temp. not over 10 °C), the sample was taken up into EtOH or into ethyleneglycolmonomethylether (depending on its polarity) and subjected immediately to either semi-preparative or analytical HPLC. This extraction procedure takes about 30 min.

For preparative work (e.g., production of reference substances or of enzyme substrates), compounds were fractionated by semi-preparative HPLC (for separation conditions, see Fig. 1), collected and immediately freeze-dried. Dried material was taken up in solvents as named above and stored at −20 °C. If necessary, a cochromatography with authentic reference substances or, in the case of contamination, a second semi-preparative HPLC was performed. Reference substances were from our laboratory stock from previous work (Sütfeld 1982, 1984; Pensl and Sütfeld 1985) or were obtained from ALDRICH-CHEMIE (Steinheim, FRG). Using this purification/storage procedure, no decomposition of thiophene derivatives could be observed over months.

3 Use of HPLC for Phytochemical Research on Thiophenes

3.1 Preliminary Experiments: HPLC of Commercially Available Reference Substances

As indicated above, the spectrum of thiophenes of an unknown natural system can be expected to be very broad with regard to the number and to the polarity of the compounds in question. To figure out the most suitable HPLC separation system, performing of preliminary experiments using standard test substances is strongly recommended. The compounds shown in Table 1 are commercially available; because of their differences in polarity, substitution, and chain lengths, they can be used as a tool for the evaluation of suitable HPL chromatographic separation conditions. As shown in Table 1, a RP18 column fractionates these compounds under isocratic conditions in a satisfactory manner. However, retention times achieved for the most apolar compounds (thiophene and bithiophene) seem to be too high; therefore, it seems to be recommendable to apply gradients

Table 1. Isocratic reverse-phase HPLC[a] of monothiophenic and of dithiophenic reference substances

Compound	Formula	Detection[b] (nm)	Rt (min)
Thiophene-2-carbonic acid	C_4H_3SCOOH	260	3.55
3-(2-Thienyl)-acrylic acid	$C_4H_3SCH{=}COOH$	235	4.26
2-Thienylacetic acid	$C_4H_3SCH_2COOH$	305	4.36
4-(2-Thienyl)-butyric acid	$C_4H_3S(CH_2)_3COOH$	245	6.46
2-(2-Thienyl)-ethanol	$C_4H_3S(CH_2)_2OH$	234	9.16
Thiophene	C_4H_4S	232	24.17
2,2′-Bithiophene	$(C_4H_3S)_2$	245	32.81

[a] Col.: LiChrosorb RP 18, 5 μ, 8 × 250 mm; Solvent system: CH_3CN/H_2O 40 : 60, isocratic; Flow: 1.8 ml min^{-1}; Press: 24 MPa.
[b] Spectrophotometrically, in the range of the absorption maxima of the respective compounds.

for further work on naturally occurring compounds as shown in the next figures and Table 2.

3.2 HPLC of Naturally Occurring Thiophenes

3.2.1 Thiophenes from *Tagetes patula* Seedlings

The analytical spectral data of the mainly accumulated thiophene derivatives of *Tagetes patula* (4-Hydroxybutinylbithiophene **1**, 4-Acetoxybutinylbithiophene **6,** Dithienylbutinene **11,** and Terthiophene **13**) are well established (Zechmeister and Sease 1947; Bohlmann and Herbst 1962; Bohlmann et al. 1964, 1973; Bohlmann and Zdero 1979). These compounds exhibit in common a broad adsorption band at about 330–350 nm; thus, in the short-wave UV at about 260 nm, they adsorb to a much lesser extent. It is therefore recommended to perform comparable HPLC at both wavelength ranges if crude plant extracts are to be examined for their contents on thiophenes as shown in Fig. 1: Compounds **2** and **7,** for example, show a stronger adsorption in the UV-B region in contrast to compounds **6** and **11.** Therefore, it can be suggested that at least **6** and **11** most probably have a thiophenic character, rather than **2** and **7** which may belong to other classes of natural products.

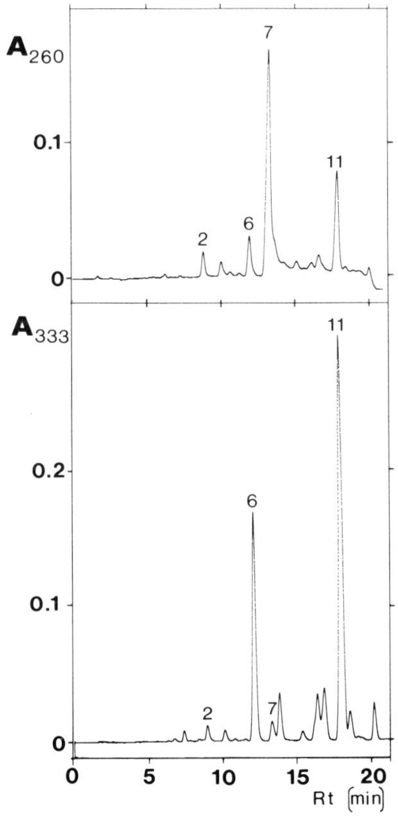

Fig. 1. HPL-chromatographic separation of EtOH/PE extractable natural compounds of a sample from *Tagetes* seedlings. Detection performed at short-wave (A_{260}, above) and at long-wave (A_{333}, below) UV. Clearly identified compounds are numbered as shown in Fig. 3. Col.: LiChrosorb RP18, 5 μ, 8 × 250 mm; Solvent system: Gradient $H_2O/CH_3CN–H_2O$-Azeotrop/CH_3CN (20 : 80 : 0) to $H_2O/CH_3CN–H_2O$-Azeotrop/CH_3CN (0/100/0) in 15 min, then 100% CH_3CN; Flow: 2.5 ml min^{-1}; Press: 16.5 MPa

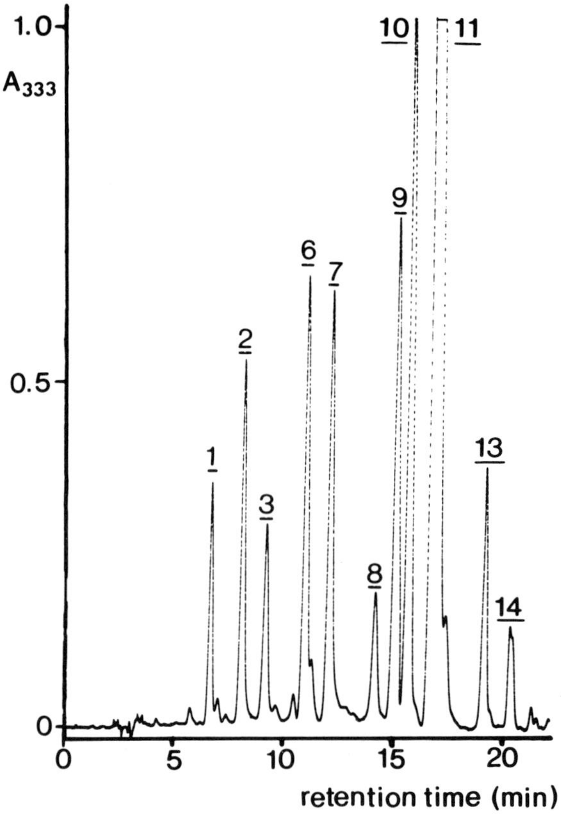

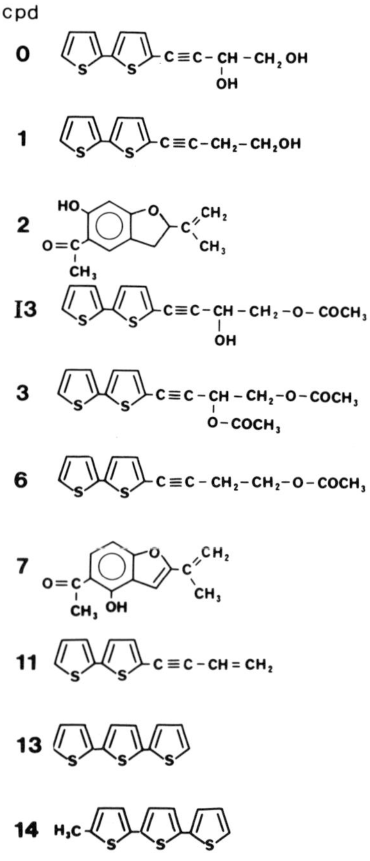

Fig. 2 Fig. 3

Fig. 2. Semi-preparative HPLC of a crude EtOH/PE extract from roots of *Tagetes* seedlings, grown for 6 days in the dark and then illuminated for 2 days. Separation conditions as shown in Fig. 1. (Pensl and Sütfeld 1985)

Fig. 3. Structures of UV-A absorbing natural compounds from *Tagetes patula,* isolated by semi-preparative HPLC. Compounds are ordered in the range of succeeding Rt values (decreasing polarity) and numbered as shown in Fig. 2. Identity of substances was established by HPL-cochromatography, GC-MS, IR-, and UV-spectroscopy as well as by enzymological work. Analytical data for structure evaluation were given by Bohlmann et al. (1973) and cited literature therein (cpds. **1, 6, 11, 13**); Bohlmann and Grenz (1970) (cpd. **2**); Downum (1984), personal communication (cpd. **14**); Pensl and Sütfeld (1985) (cpds. **0, 3**); and by Sütfeld et al. (1985) (cpd. **7**)

In this way, using analytical and semi-preparative HPLC (see Fig. 2) with subsequent UV-, MS- and NMR spectroscopy, almost the entire spectrum of UV-absorbing natural products of *Tagetes patula* seedlings could be elucidated (Fig. 3; Sütfeld and Towers 1982; Sütfeld 1982; Downum and Towers 1983; Sütfeld 1984; Pensl and Sütfeld 1985). Compounds **2** and **7** proved be no-thiophenic as suggested above (Fig. 1) whereas compound **14** seems to be a methylated terthiophene (Downum 1984, personal communication). Additionally, two minor peaks, corresponding to the structures of compounds **0** and **13**, could be identified by enzymatical production of reference substances (Pensl and Sütfeld 1985).

3.2.2 Thiophenes from Other Plant Sources

The well-established system, *Tagetes patula* seedlings, makes it easy to look for thiophene contents in other "unknown" plant systems using comparative HPLC.

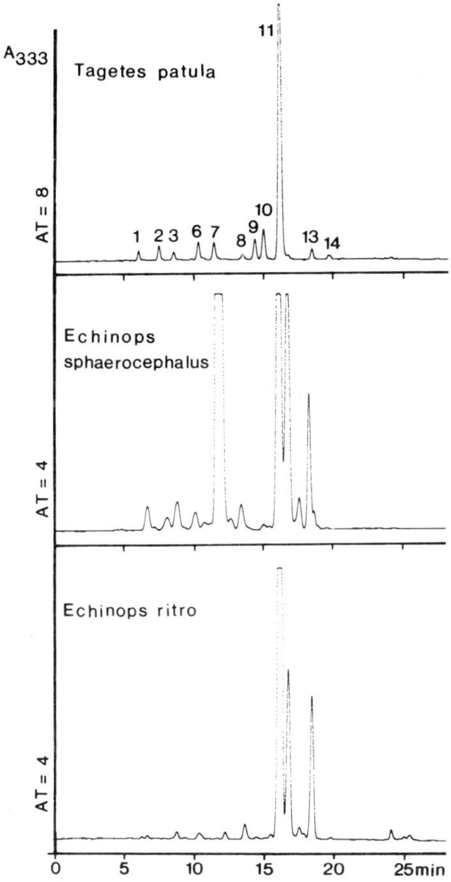

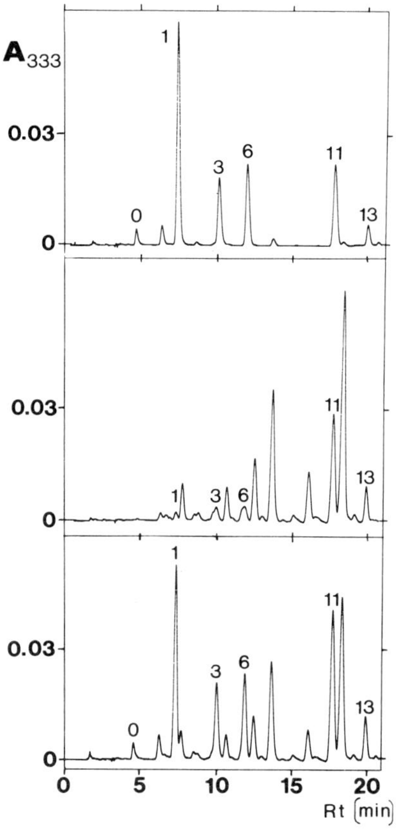

Fig. 4 Fig. 5

Fig. 4. Comparative search for thiophenes in different plant systems by HPLC. *Above* "Standard" sample (extract from roots of *Tagetes* seedlings). Compounds numbered as shown in Figs. 2 and 3. *Middle and below* Samples from roots of adult *Echinops sphaerocephalus* and of *E. ritro* plants. For better comparison, the respective absorbance range (full scale = 2.0 A_{333}) was attenuated (AT) eightfold or fourfold, respectively. Separation conditions as shown in Fig. 1

Fig. 5. Comparative cochromatography of thiophene derivatives. *Above* Mixture of purified thiophenes from *Tagetes* (numbered as shown in Fig. 3). *Middle* EtOH/PE extract from hypocotyls of *Echinops ritro,* grown for 7 days in the dark and then illuminated for 5 days. *Below* Result of cochromatography of both the samples. Separation conditions as shown in Fig. 1

Table 2. Qualitative/quantitative cochromatography[a] of reference substances with unknown samples

Compound no.	0		1		3		6		11		13	
	Rt	A.U.[b]	Rt	A.U.	Rt	A.U.	Rt	A.U.	Rt	A.U.	Rt	A.U.
Reference substances	4.56	1.148	7.00	31.683	9.88	5.292	11.89	11.384	17.82	17.201	20.06	3.044
Echinops ritro (Hypocotyls)	n.d.[c]	n.d.	n.d.	n.d.	9.98	1.675	11.80	2.112	17.85	11.806	20.04	3.007
Cochromatography[d]	4.56	1.148	7.00	31.683	9.98	6.967	11.85	13.496	17.84	29.007	20.05	6.051
[e]	4.58	1.177	7.02	32.577	9.95	6.794	11.82	12.998	17.83	29.355	20.03	6.241
Deviation[f]		2.5		2.7		2.5		3.8		1.1		3.0

[a] Col.: LiChrosorb RP 18, 5 μ, 8 × 250 nm; Solvent system: Gradient H_2O/CH_3CN–H_2O-Azeotrop/CH_3CN (20 : 80 : 0) to H_2O/CH_3CN–H_2O-Azeotrop/CH_3CN (0 : 100 : 0) in 15 min, then 100% CH_3CN; Flow: 2.5 ml min^{-1}; Press: 24 MPa.
[b] A.U. = Absorption Units (A_{333}), measured by peak integration.
[c] n.d. = not detectable.
[d] Expected values by mean.
[e] Measured values.
[f] Deviation (%) from measured values.

Figure 4 shows an example. Above, a strongly attenuated semi-preparative HPL-chromatogram of a "standard" *Tagetes* extract and below, two samples of extracts from *Echinops* plants which are suggested to contain several thiophene derivatives (Bohlmann et al. 1965, 1973). From the results obtained, it can be roughly estimated that both the *Echinops* species in fact contain some of the same thiophenic compounds of *Tagetes*. The evidence becomes much clearer by quantitative cochromatography of distinct reference substances isolated from *Tagetes* (Fig. 5; Table 2). By comparison of the Rt values and of the integrated data, it can be strongly assumed that at least compounds **3, 6, 11,** and **13,** and, perhaps, **0** and **1,** too, are most probably present in the *Echinops* extracts. This result does not exclude the necessity of further spectroscopical work regarding structure determination; however, it is a certain hint for the occurrence of these compounds in question in this "unknown" plant material.

4 Use of HPLC for Enzymatic Research on Thiophenes

Performing enzymatic work on thiophene derivatives, an isocratic solvent system proved sufficient for most purposes (Sütfeld and Towers 1982; Pensl and Sütfeld

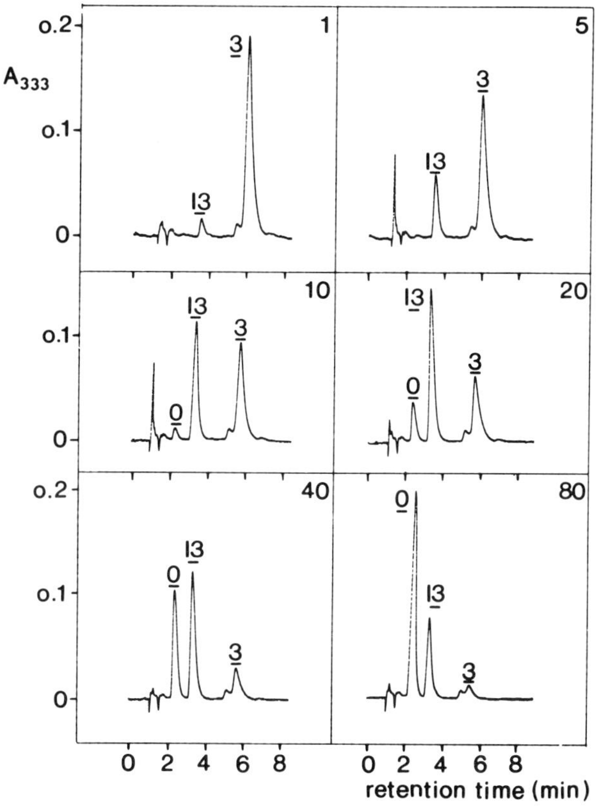

Fig. 6. Application of HPLC for enzymic research on thiophenes. Substrate: Cpd. **3.** Intermediate product: Cpd. **13**. End product: Cpd. **0.** *Numbers in the upper right corner* represent the time of enzyme assay incubation. Enzyme source: Crude extracts from roots of *Tagetes patula,* grown for 2 days in the dark and then illuminated for 7 days. Separation conditions: Col.: Spherisorb ODSII, 5 μ, 4 × 250 mm; Solvent system: H_2O/CH_3CN (25:75), isocratic; Flow: 1.6 ml min^{-1}; Press: 14.4 MPa. (Pensl and Sütfeld 1985)

1985). Figure 6 shows that also complicated assays like two-enzyme reactions can easily be analyzed by observation of substrate decreasement and of product formation over the respective period of incubation time. By use of suitable guard columns, samples can be safely applied to HPLC without previous (acid) protein precipitation. This avoids experimental errors which are possible by acid treatment of the samples (e.g., if hydrolyzable compounds are involved in the enzyme reactions).

5 Conclusions

HPLC proved to be a fast, powerful, and careful tool for exact qualitative and quantitative analyses of naturally occurring thiophene derivatives. Deep-temperature extraction, connected with this technique, avoids decomposition of these fairly instable plant pigments. Because of the chemical characters of most of the investigated compounds, relatively simple solvent mixtures or gradients are required for yielding satisfactory separations. CH_3CN/H_2O mixtures with a high content of CH_3CN seem to be most suitable. Quantitative HPL cochromatography using authentic reference substances appears to be one of the easiest ways for the elucidation of unknown thiophene-containing samples from any plant source. Quantitation of enzyme assays with thiophenic substrates is likewise easily possible.

Acknowledgments. The loan of HPLC thru-flow spectrophotometers for these experiments is gratefully acknowledged to the companies KRATOS (Ramsey, NJ, USA) and KNAUER (Bad Homburg v.d.H., FRG). I wish to thank the Deutsche Forschungsgemeinschaft for financial support in the scope of the research group Sekundäre Naturstoffe/Zellkulturen.

References

Arnason T, Towers GHN, Philogene BJR, Lambert JDH (1983) The role of natural photosensitizers in plant resistance to insects. ACS Symp Ser 208:139–151

Bohlmann F, Grenz M (1970) Neue Isopentenyl-acetophenon-Derivate aus *Helianthella uniflora*. Chem Ber 103:90–95

Bohlmann F, Herbst P (1962) Über die Inhaltsstoffe von *Tagetes*-Arten. Chem Ber 95:2945–2955

Bohlmann F, Zdero C (1979) Über dimere Terpenketone aus *Tagetes gracilis*. Phytochemistry (Oxf) 18:341–343

Bohlmann F, Kleine KM, Arndt C (1964) Über natürlich vorkommende Thiophenacetylenverbindungen. Chem Ber 97:2125–2134

Bohlmann F, Arndt C, Kleine KM, Bornowski H (1965) Die Acetylenverbindungen der Gattung *Echinops*. Chem Ber 98:155–163

Bohlmann F, Burkhardt T, Zdero C (1973) Naturally occurring acetylenes. Academic Press, New York

Downum KR, Towers GHN (1983) Analysis of thiophenes in the tageteae (asteraceae) by hplc. J Natl Prod 46:98–103

Pensl R, Sütfeld R (1985) Occurrence of 3,4-diacetoxybutinylbithiophene in *Tagetes patula* and its enzymatic conversion. Z Naturforsch Sect C Biosci 40:3–7

Sørensen NA (1977) Polyacetylenes and conversation of chemical characters in the compositae. In: Heywood VH, Harborne JB, Turner BL (eds) Biology and chemistry of the compositae. Academic Press, London, pp 385–409

Sütfeld R (1982) Distribution of thiophene derivatives in different organs of *Tagetes patula* seedlings grown under various conditions. Planta (Berl) 156:536–440

Sütfeld R (1984) Phytochemische und enzymologische Untersuchungen zum Metabolismus von sekundären Naturstoffen in *Tagetes*-Keimlingen. Habilitation Thesis, Westf Wilh Univ, Münster

Sütfeld R, Towers GHN (1982) 5-(4-Acetoxy-1-butinyl)-2,2′-bithiophene: acetate esterase from *Tagetes patula*. Phytochemistry (Oxf) 21:277–279

Sütfeld R, Balza F, Towers GHN (1985) A new benzofuran from *Tagetes* seedlings. Phytochemistry (Oxf) 24:876–877

Towers GHN (1980) Photosensitizers from plants and their photodynamic action. In: Reinhold L, Harborne JB, Swain T (eds) Prog Phytochem, vol 6. Pergamon, Oxford, pp 183–202

Zechmeister L, Sease J (1947) A blue fluorescing compound, terthienyl, isolated from marigolds. J Am Chem Soc 69:273

High Performance Liquid Chromatography of Ascorbic Acid

J. P. F. G. Helsper

1 Introduction

The relation between the consumption of fresh fruits and vegetables and the prevention of scurvy has been known for several centuries. The chemical structure of the biologically active compound involved was completely elucidated in 1933 by Herbert and coworkers and designated L-threo-hex-2-enono-1,4-lactone. Trivial names for this compound are L-ascorbic acid (AA*) or vitamin C. More recent investigations indicate that there is a very wide spectrum of medical and health applications for this vitamin.

Due to the lack of the enzyme L-gulonolactone oxidase, humans, as well as primates, guinea pigs, several fruit-eating bats and highly evolved Passeriform birds are not capable of synthesizing AA* from carbohydrate precursors. The fact that higher plants are the major source for this vitamin created the need for an easy, fast, and sensitive procedure to determine its content in plant tissues and food products. High-performance liquid chromatography meets these demands. This technique has been used in both fundamental and applied research on AA metabolism and biosynthesis.

HPLC equipment for analysis of AA in plant extracts is commercially available. Solvent mixing equipment is usually not required, since almost all analyses reported to date are performed isocratically. Due to the chemical lability of AA, special consideration has to be given to the preparation of plant extracts. Section 2 will deal with sample preparation, Sect. 3 describes combinations of stationary and mobile phases that can be used for specific research goals and Sect. 4 gives a survey of detection systems, applied for the quantitative assay of AA.

2 Sample Preparation

Since AA is an easily oxidizable compound, precautions have to be made to safeguard its reduced form. Oxidation by molecular oxygen or metal ions, resulting at first instance in the formation of DHAA, can be diminished by homogenization at low temperature in acidic solvents (Rose and Nahrwold 1982). The low pH stabilizes the 1,4-lactone structure of AA and thus lowers its sensitivity toward oxidation (Omaye et al. 1979).

Oxalic acid (0.1–2%), metaphosphoric acid (1–5%), perchloric acid (0.05 M), acetic acid (1–10%), or combinations of these are frequently used as acidifiers.

No additional buffer salts are necessary. Acidic solutions of high ionic strength, however, can damage the HPLC equipment, especially the column material, so that dilution of the sample before injection is sometimes required. Oxidation by metal ions can be reduced by adding chelators such as EDTA or EGTA (about 1 mM) to the homogenization buffer. Oxalic acid acts both as a chelator and an acidifier.

A reducing state of the extraction medium can also be achieved by the addition of reducing agents, such as DTT, cysteine (both at approximately 0.1%), GSH (1–5 mM) or by bubbling with hydrogen sulfide. These treatments are known to reduce endogenous DHAA to AA (Doner and Hicks 1981; Okahama 1981; Farber et al. 1983), so that the measured AA level may become higher than that in the original tissue.

3 Stationary and Mobile Phases

The choice for a specific combination of column-filling material (stationary phase) and eluens (mobile phase) for HPLC of AA will greatly depend on the aim of analysis. For fast quantitative assays of AA, e.g., in fruit juices, plant extracts, and food products, a reliable sample preparation method (see Sect. 2), a high sampling rate and a sensitive and selective detection system, e.g., electrochemical detection or UV absorbtion (see Sect. 4), are the most important requirements.

Several modes of HPLC have been used for these AA assays, such as anion exchange (Pachla and Kissinger 1979; Doner and Hicks 1981), cation exchange (Grün and Loewus 1984), and ion-pair reversed-phase (Sood et al. 1976; Moledina and Flinck 1982). Actually any combination of stationary and mobile phase is appropriate, as long as the retention time for AA is relatively short. In performing routine AA assays it is relevant to know that, after storage of the column in methanol, the first injections might not give 100% recovery for AA (Pachla and Kissinger 1976, 1979).

Other applications of HPLC, such as separation of AA from metabolically related products, DHAA and reducing agents, and the resolution of isomeric forms of ascorbic acid require more specific combinations of stationary and mobile phases. These are discussed below in more detail.

3.1 Separation of Ascorbic Acid from Dehydroascorbic Acid and from Reducing Agents

Reducing agents are often added during or shortly after preparation of plant homogenates to prevent oxidation of AA (see Sect. 2). These agents can reduce endogenous DHAA to AA, thus increasing the level of AA before it is assayed. Reducing agents can also interfere with the detection method itself, since most of the detection methods are based on oxidation-reduction reactions with AA. This latter method of interference by reducing agents can be circumvented by separation from AA prior to its detection. Sybilska et al. (1984) used a reversed-phase system for this purpose with 0.067 M phosphate-buffered 0.25% oxalic acid (pH 2.9) as the mobile phase (flow rate 1.0 ml min^{-1}) and 10 µm LiChrosorb RP-18 as the stationary phase and achieved fast separation of AA from cysteine and GSH. Separation of AA from DTT is reported by Farber et al. (1983) with a Whatman Partisil 10 SAX stationary phase and 7 mM KCl/7 mM KH_2PO_4 (pH 4.0) as the mobile phase.

Dehydroascorbate can be separated from AA with the same combination. The retention times for DHAA and DTT are about 2 and 3 min, respectively. AA itself is retained by the column and is released 1 min after introduction of a linear gradient of the 7 mM KCl/KH_2PO_4 eluent and 500 mM KCl/250 mM KH_2PO_4 (pH 5.0).

Several other stationary and mobile phases have been used to achieve resolution of AA and DHAA. For HPLC in the anion-exchange mode phosphate-buffered, aqueous acetonitril solutions were used as the mobile phases in combination with a Zorbax Analytical column (Doner and Hicks 1981), LiChrosorb-NH_2 (Rose and Nahrwold 1981, 1982) or μBondapak C18 (Wimalassiri and Wills 1983) as the stationary phases. μBondapak C18 columns are also used for ion-pair reversed-phase HPLC of AA and DHAA. Finley and Duang (1981) use tri-n-butylamine (0.7 ml l^{-1}) in 40 mM phosphate buffer (pH 4.3) as the mobile phase, while Keating and Haddad (1982) elute with aqueous methanol, containing 0.25 mM hexadecyltrimethylammoniumbromide or 40 mM tridecylammoniumformate as ion-pair reagents. Resolution of AA and DHAA is usually very good. With the latter ion-pair reversed phase system, AA and DHAA have retention times of 1.5 and 6 min, respectively, at a flow rate of 2.0 ml min^{-1}.

3.2 Separation of Ascorbic Acid and Metabolically Related Compounds

A number of neutral sugars and their derivatives, such as D-glucose, D-galactose, L-gulono-γ-lactone and L-galactono-γ-lactone, have been postulated as possible intermediates in the biosynthetic pathway for AA in higher plants. To date this biosynthetic pathway is still an unresolved problem (Loewus and Helsper 1982). During the metabolism of AA, the first oxidation product is DHAA. Further oxidation leads to the formation of several organic acids, such as L-idonate, 2-keto-L-idonate (Saito and Kasai 1982, 1984) L-tartrate, oxalate, L-threonate and glycerate (Helsper and Loewus 1982; Loewus and Helsper 1982). Part of the six-carbon chain of AA is recycled into products of hexose phosphate metabolism.

HPLC of these compounds, with the exception of DHAA, has seldom been reported. Grün et al. (1983) attempted to separate most of these metabolites from AA and from each other on a strong cation exchange system with the organic acid column HPX-87 (Bio-Rad Laboratories, Richmond, Cal. USA) as the stationary phase and 0.1 M formic acid (40 °C) as the mobile phase. Detection with refractive index showed poor or no resolution. For this purpose classical column ion-exchange chromatography as described by Wagner and Loewus (1973) seems more appropriate. HPLC with the above system has been successfully applied for additional identification and purification, in combination with column ion-exchange chromatography, of metabolic products of AA such as L-tartrate (Grün et al. 1982), glycolaldehyde and L-malate (Helsper et al. 1981).

3.3 Separation of Isomeric Forms of Ascorbic Acid

Due to the assymetric carbons 4 and 5 in ascorbic acid, four isomers are possible, the D- and L-enantiomers of ascorbic acid and erythorbic acid. Naturally occurring ascorbate has the L-configuration. Other trivial names for EA are isoascorbic acid or araboascorbic acid. Separation of ascorbate from erythorbate has been achieved by several authors using HPLC in the reversed-phase, ion-pair reversed-phase or cation exchange mode. Ascorbic acid isomers can be detected with the techniques, described for naturally occurring L-ascorbic acid (see next section).

For reversed-phase HPLC phosphate-buffered aqueous acetonitril solutions of varying pH are used as the mobile phases and LiChrosorb-NH_2 (Bui-Nguyên 1980), Zorbax-NH_2 (Doner and Hicks 1981) or a 10 µm Carbohydrate column from Waters Associates (Geigert et al. 1981) as the stationary phases. Retention times for ascorbic acid are about 15% longer than for erythorbic acid.

As ion-pair reagents 5 mM cetyltrimethylammonium (Coustard and Suraud 1981) and 1 mM decylamine (Tsao and Salimi 1982) have been added to mobile phases consisting of aqueous methanol, buffered with 50 mM KH_2PO_4/50 mM K_2HPO_4 and 40 mM acetate (5 ml glacial acetic acid + 1.11 g anhydrous sodium acetate per liter) respectively. LiChrosorb RP-18 (Coustard and Suraud 1981) and 5 µm ultrasphere ODS (Altex) (Tsao and Salimi 1982) are used as the stationary phases. With ion-pair reversed phase HPLC ascorbic acid elutes about 15% earlier than erythorbic acid. With the strong cation-exchange system described in Sect. 3.2 (Grün et al. 1983), ascorbic acid elutes 6% later than erythorbic acid.

4 Detection Techniques for Ascorbic Acid

Techniques used for the assay of AA are based on colorimetry, fluorometry, refractive index, UV absorbtion and amperometry. Their selectivity and detection limits are listed in Table 1. Radiation detectors, which can be directly connected

Table 1. Selectivity and sensitivity of several detection methods used in HPLC of AA

Detection method	Selectivity	Detection limit	Application in HPLC of AA
Colorimetry	Poor	µg	Control for other detection methods
Refractometry	Poor	10 µg	Purification and metabolic studies
UV spectrometry	High	ng	Various applications
Fluorometry	High	ng	Detection of AA and DHAA
Amperometry	Very high	10 pg	Various applications

to the column, are also commercially available for radiotracer studies. Chemoluminescence has been applied for detection of DHAA (Veazy and Nieman 1980). The choice of assay procedure depends on the demands for sensitivity, selectivity, and recovery. When only the AA level in a certain tissue extract is to be determined, a highly selective and sensitive method is required. In contrast, selectivity is not desirable when components other than AA also have to be detected, e.g., in purification procedures. Colorimetric, fluorometric, and amperometric assays involve oxidation of AA and therefore always lead to less than 100% recovery.

The most commonly used dyes for colorimetric methods are 2,6-dichlorophenolindophenol, 2,4-dinitrophenylhydrazine, and α,α′-dipyridyl. The procedures in which these compounds are used for AA detection are described by Omaye et al. (1979). They are also applied for AA determination in tissue extracts without previous chromatographic purification. Due to the accompanying loss of material, the insensitivity, and low selectivity in comparison to other available techniques, colorimetric assays are not routinely used in HPLC analyses, but merely as a control for other detection methods.

Fluorometric assay requires oxidation of AA to DHAA. Therefore this technique has also been used for the assay of DHAA. This latter compound can undergo condensation with o-phenylenediamine to form the fluorescent quinoxaline derivative, which can be quantified by fluorometry at 348 nm. Fluorometric assay of AA and DHAA has been described by Omaye et al. (1979) and more recently by Baker et al. (1983). A high sensitivity can be obtained when the quinoxaline derivative is synthesized prior to injection and assayed directly after elution from the column (Keating and Haddad 1982; Speek et al. 1984). Speek et al. (1984) could detect as little as 4 pmol in this way.

Refractive index is not the most sensitive detection method for AA, the detection limit being about 10 μg. Due to its low selectivity however, it is very appropriate to detect compounds other than AA in HPLC analyses. This may be relevant in purification procedures or studies on AA metabolism (see Sect. 3.2) and derivatization. Refractive index, which can be directly coupled to HPLC equipment, is commercially available. Thus separation and detection are almost simultaneous and the recovery is 100%.

UV absorbtion measurement is much more sensitive and selective for AA than refractive index detection. Farber et al. (1983) detected as little as 4 ng AA at 254 nm. Comparable detection limits are observed for isomers of naturally occurring L-ascorbic acid (Doner and Hicks 1981; Geigert et al. 1981). Some compounds related to AA metabolism can also be quantified by UV absorbtion, mostly at lower sensitivity. Wimalassiri and Wills (1983) mention a detection limit of 100 ng for DHAA at 214 nm. UV absorbtion detectors can be built in as a component of the HPLC equipment with the advantage of rapid detection and complete recovery.

Amperometric or electrochemical detection is certainly the most sensitive and most selective assay method for AA. With a dropping mercury electrode set at 500 mV vs. Ag/AgCl Sybilska et al. (1984) reported a detection limit of 5 ng. Sensitivities beyond the nanogram level are obtained with glassy carbon flow cells, e.g., of the LC- or CPW-series from Bioanalytical Systems (West Lafayette, Ind., USA) (Tsao and Salimi 1982; Grün and Loewus 1983). With an ECP-1 glassy carbon electrode from Kotaki (Funabashi, Chiba, Japan), set at 800 mV vs. Ag/AgCl, quantitaties of 10 pg AA have been detected in human serum (Iriyama et al. 1984) and various rat tissues (Aoki et al. 1984).

References

Aoki T, Yoshiura M, Iwamoto T, Iriyama K (1984) Postmortem changes in uric acid in various rat tissues: determination of uric acid by reversed-phase high-performance liquid chromatography with electrochemical detection. Anal Biochem 143:113–118

Baker JK, Kapeghian J, Verlangieri A (1983) Determination of ascorbic acid and dehydroascorbic acid in blood plasma samples. J Liq Chromatogr 6:1319–1332

Bui-Nguyên MH (1980) Application of high-performance liquid chromatography to the separation of ascorbic acid from isoascorbic acid. J Chromatogr 196:163–165

Coustard JM, Suraud G (1981) Séparation des acides ascorbique et isoascorbique par chromatographie des paires d'ion sur phase inverse. J Chromatogr 219:338–342

Doner LW, Hicks HB (1981) High-performance liquid chromatographic separation of ascorbic acid, erythorbic acid, dehydroascorbic acid, diketogulonic acid and diketogluconic acid. Anal Biochem 115:225–230

Farber AM, Kamengiser S, Stahl R, Liebes L, Silber R (1983) A specific high-performance liquid chromatography assay for dehydroascorbic acid shows an increased content in chronic lymphocytic leukemia lymphocytes. Anal Biochem 134:355–360

Finley JW, Duang E (1981) Resolution of ascorbic, dehydroascorbic and diketogluconic acids by paired-ion reversed-phase chromatography. J Chromatogr 207:449–453

Geigert J, Hiramo DS, Neidleman SL (1981) High-performance liquid chromatographic method for the determination of L-ascorbic acid and D-isoascorbic acid. J Chromatogr 206:396–399

Grün M, Loewus FA (1983) Determination of L-ascorbic acid in algae by high-performance liquid chromatography on strong cation-exchange resin with electrochemical detection. Anal Biochem 130:191–198

Grün M, Renstrøm B, Loewus FA (1982) Loss of hydrogen from carbon 5 of D-glucose during conversion of D-[5-^{3}H, 6-^{14}C] glucose to L-ascorbic acid in *Pelargonium crispum* (L.) L'Hér. Plant Physiol 70:1233–1235

Helsper JPFG, Loewus FA (1982) Metabolism of L-threonic acid in *Rumex x acutus* and *Pelargonium crispum* (L.) L'Her. Plant Physiol (Bethesda) 69:1365–1368

Helsper JPFG, Saito K, Loewus FA (1981) Biosynthesis and metabolism of L-ascorbic acid in Virginia Creeper (*Parthenocissus quinquefolia* L.). Planta (Berl) 152:171–176

Herbert RW, Hirst EL, Percival EGV, Reynolds RJW, Smith F (1933) The constitution of ascorbic acid. J Chem Soc (Lond) 1270–1290

Iriyama K, Yoshiura M, Iwamoto T, Ozaki Y (1984) Simultaneous determination of uric and ascorbic acids in human serum by reversed-phase high-performance liquid chromatography with electrochemical detection. Anal Biochem 141:238–243

Keating RW, Haddad PR (1982) Simultaneous determination of ascorbic acid and dehydroascorbic acid by reversed-phase ion-pair high performance liquid chromatography with pre-column derivatization. J Chromatogr 245:249–255

Loewus FA, Helsper JPFG (1982) Metabolism of L-ascorbic acid in plants. In: Seib PA, Tolbert BM (eds) Ascorbic acid: chemistry, metabolism, and uses. Adv Chem Ser Am Chem Soc, vol 200. Washington DC, pp 249–261

Moledina KH, Flinck JM (1982) Determination of ascorbic acid in plant food products by high-performance liquid chromatography. Lebensm Wiss Technol 15:351–358

Okamura M (1981) An improved method for determination of ascorbic acid in plasma and urine by the α,α'-dipyridyl method. Vitamins (Kyoto) 55:495–508

Omaye ST, Turnbull JD, Sauberlich HE (1979) Selected methods for the determination of ascorbic acid in animal cells, tissues and fluids. Methods Enzymol 62:3–14

Pachla LA, Kissinger PT (1976) Determination of ascorbic acid in foodstuffs, pharmaceuticals, and body fluids by liquid chromatography with electrochemical detection. Anal Biochem 48:364–367

Pachla LA, Kissinger PT (1979) Analysis of ascorbic acid by liquid chromatography with amperometric detection. Methods Enzymol 62:15–24

Rose RC, Nahrwold DL (1981) Quantitative analysis of ascorbic acid and dehydroascorbic acid by high-performance liquid chromatography. Anal Biochem 114:140–145

Rose RC, Nahrwold DL (1982) Ascorbic acid analysis in body fluids. Anal Biochem 123:389–393

Saito K, Kasai Z (1982) Conversion of L-ascorbic acid to L-idonic acid, L-idono-γ-lactone, and 2-keto-L-idonic acid in slices of immature grapes. Plant Cell Physiol 23:499–507

Saito K, Kasai Z (1984) Synthesis of L-(+)-tartaric acid from L-ascorbic acid via 5-keto-D-gluconic acid in grape. Plant Physiol (Bethesda) 76:170–174

Sood SP, Sartori LE, Wittmer DP, Haney WG (1976) High-pressure liquid chromatographic determination of ascorbic acid in selected foods and multivitamin products. Anal Chem 48:796–798

Speek AJ, Schrijver J, Schreurs W (1984) Fluorimetric determination of total vitamin C in whole blood by high-performance liquid chromatography with precolumn derivatization. J Chromatogr Biomed Appl 305:53–60

Sybilska D, Duszczyk K, Przasnyski M (1984) Determination of ascorbic acid in its mixtures with thiols by reversed-phase high-performance liquid chromatography with electrochemical detection using a DC-TAST polarograph. J Chromatogr 298:352–355

Tsao CS, Salimi SL (1982) Differential determination of L-ascorbic acid and D-isoascorbic acid by reversed-phase high-performance liquid chromatography with electrochemical detection. J Chromatogr 245:355–358

Veazy RL, Nieman TA (1980) Chemiluminescence high-performance liquid chromatographic detector applied to ascorbic acid determinations. J Chromatogr 200:153–162

Wagner G, Loewus F (1973) The biosynthesis of L-(+)-tartaric acid in *Pelargonium crispum*. Plant Physiol (Bethesda) 52:651–654

Wimalassiri P, Wills RBH (1983) Simultaneous analysis of ascorbic acid and dehydroascorbic acid in fruit and vegetables by high-performance liquid chromatography. J Chromatogr 256:368–371

High Performance Liquid Chromatography of Phytoalexins

R. N. STRANGE

1 Introduction

Phytoalexins have been defined as "low molecular weight, antimicrobial compounds that are both synthesized by and accumulated in plants after exposure to microorganisms" (Paxton 1981). They have been studied from many points of view; as natural products, as chemotaxonomic markers, as antimicrobial agents, and as the indirect products of gene derepression. Investigations of the mechanisms by which their accumulation is triggered (or, to use the accepted terminology, elicited) are numerous and considerable attention has also been paid to both biosynthesis and degradation. Above all, it is the putative role of phytoalexins in the defence of plants against microbial attack which provides the most important stimulus for their study. Recent work in this area has been encouraging and before long we may find plant breeders, in their quest for resistance, wishing to select lines with a high capacity for phytoalexin synthesis. If this does come about the requirement for speedy methods of qualitative and quantitative analysis will be enormous.

The technique of HPLC is ideally suited to both the qualitative and quantitative study of phytoalexins. Its high resolving power allows their separation from complex plant extracts and in some instances this has led to the discovery of novel phytoalexins, not previously separated by either TLC or GLC (Porter et al. 1979), while the speed and mildness of the separation process itself makes for good recovery. In addition, the high extinction coefficients of many phytoalexins in the ultraviolet region of the spectrum enables the experimenter to monitor column effluents with a flow through UV detector and obtain accurate quantitative analysis of the compounds in the microgram or submicrogram range.

It is not the aim of this chapter to give a general account of the principles and practice of HPLC, since this may be obtained from several books on the subject, but rather to dwell on those aspects of methodology which, it is hoped, will help the experimenter to prepare and quantitatively to analyse phytoalexins successfully. First, features of HPLC instrumentation which are of particular value for phytoalexin studies are reviewed. This is followed by a section on preparative HPLC of phytoalexins, since an early requirement in any research with these compounds is to isolate them so that their chemical structures and antimicrobial properties may be ascertained. The section is divided into two, with the first part dealing with the principles underlying the preparation of pure samples and the second with practical examples from the literature. The same approach is adopted with the next section in which the subject is the quantitative analysis of phytoalexins. Here it is hoped to provide the reader with sufficient information to enable him to optimize his procedures so that reliable data may be obtained with minimal expenditure of time, solvents and effort. Finally, consideration is given to the contribution that HPLC could make to the solution of some of the outstanding problems of phytoalexin research.

2 Instrumentation

There are now many sophisticated HPLC instruments on the market and, given the funds, experimenters will have a difficult task in choosing amongst them. For preparative work with large columns (e.g. 30 × 6 cm i.d.), a pump capable of delivering at least 100 ml min^{-1} is required but for analytical work columns rarely exceed 25 × 0.5 cm i.d. and much lower flow rates (1–2.5 ml min^{-1}) are used. Most pumps are capable of working against a back pressure of up to 5000 psi., but in practice pressure seldom exceed 1000 psi.

With large-scale preparative HPLC, samples are introduced by stopping the solvent flow and injecting at low pressure by means of a syringe. Smaller samples used for semi-preparative or analytical work are usually introduced via a valve. The sample is loaded at low pressure into a sample loop which is attached to the valve and the contents of the loop are integrated with the solvent flowing onto the column by movement of a lever. Sample loops are interchangeable and may be obtained in a wide variety of sizes from a few microlitres to 2 ml. For routine quantitative analysis of many samples an autosampler represents a considerable saving in labour. In addition the instrument may be left to run analyses unattended for long periods (e.g. overnight).

When there are many samples to analyze, reduction of analysis time to the minimum is crucial. Often plant extracts eluted in a single solvent system (i.e. isocratically) give a number of poorly resolved peaks early in the run and one or more late peaks. The latter represent not only loss of time but increased use of solvent. In order to combat this problem, a number of manufacturers have introduced gradient elution systems. These allow complex gradients to be generated in order to resolve all peaks within a minimum time. The system illustrated in Fig. 1, for example, gives the opportunity for the experimenter to mix two solvents in any proportion and to programme convex, linear or concave gradients between varying proportions of them at up to ten points during a single run.

A UV monitor is a convenient means of detecting many phytoalexins in the effluent from HPLC columns, since they generally absorb strongly in the ultraviolet range of the spectrum. Some models operate at a fixed wavelength, but this may be varied by the insertion of filters. Others offer variable wavelength. Although absorption of light at a specific wavelength and correspondence of retention time with a known standard are frequently used to identify compounds, a suspicion remains that other materials may be co-eluting with the compounds of interest. Various methods have been used to overcome this criticism. One is to monitor the effluent at two wavelengths and to plot a wavelength ratio. If a peak is given by a single compound a "square wave" will result but this will be distorted

Fig. 1. A modern HPLC with, from left to right, three units on top of each other, automatic injection controller, chromatography interface and diode array detector; automatic injector on a column oven; solvent tray on two pumps, video display screen and keyboard; printer and computer (Courtesy of Pye Unicam Ltd., York Street, Cambridge CB1 2PX, UK)

by contaminants. Another method is to stop the flow of the chromatogram and record the spectrum of the compound. Most sophisticated of all (and most expensive!) is the diode array detector which allows continuous spectral recording during elution. The model illustrated in Fig. 1 can be used to monitor in both the UV and the visible spectrum. Purity of peaks may be checked by matching spectra early and late in their elution and the spectra may also be matched with those of authentic compounds stored in the memory as an aid in identification.

In some instances it may be worth considering a fluorescence monitor. With favourable compounds sensitivity may be 100 times greater than UV absorption. In addition, considerable selectivity may be achieved by careful choice of excitation and emission wavelengths.

For compounds without appreciable light absorption in the UV range a refractometer may be used, but it is generally not very sensitive. This insensitivity is an advantage in large-scale preparative work, but a disadvantage of refractometry is that it cannot normally be used in conjuction with gradient elution. A more recently developed "universal" detector which is compatible with gradient elution is the mass detector (Fig. 2). Here the effluent from an HPLC column is nebulized and the solute particles detected by light scattering.

There is a considerable range of recorders which handle data from monitors. These vary in sophistication from a single pen chart recorder to computers. The

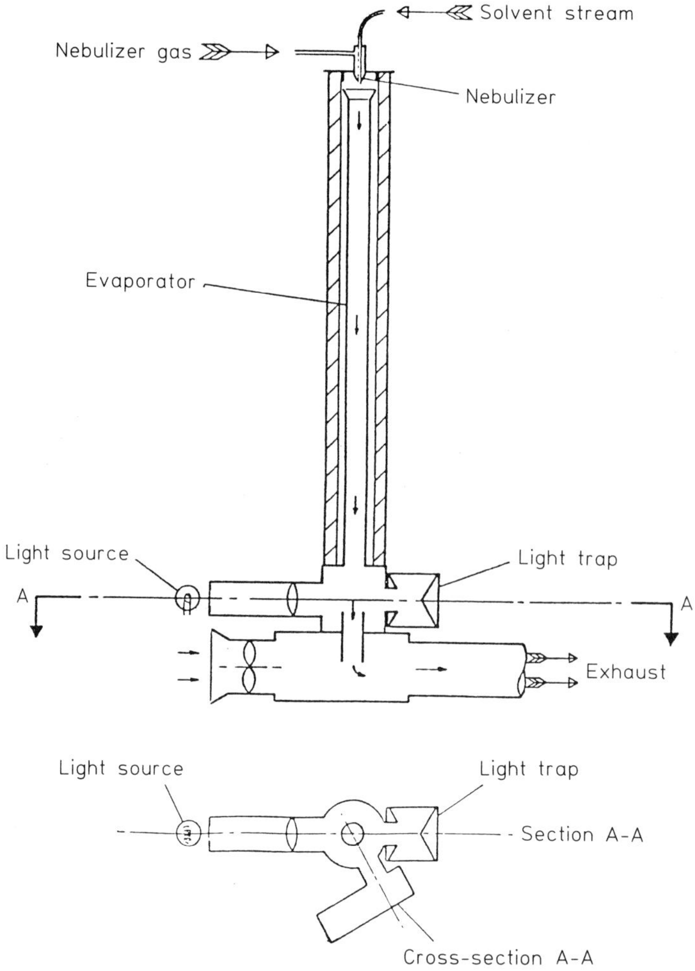

Fig. 2. A mass detector (diagrammatic) which may be used for quantitative analysis of compounds without chromaphores. Effluent from an HPLC column is fed into the warm evaporator together with a nebulizer gas such as clean and dry nitrogen. Solute particles are detected by light scattering as they pass through a light path. The refracted light is detected by a photomultiplier which lies at an angle of 120° to the incident beam and the signal amplified and displayed on a strip chart recorder. (Courtesy of Applied Chromatography Systems, Luton, Beds UK)

latter allow results to be expressed directly in absolute amounts, amounts relative to an internal or external standard or to standard addition. The computer may be used in conjunction with a data store and microprocessor interface to control peripherals such as the autosampler, elution gradients, column jacket temperature and – a matter of considerable importance if waste is to be avoided – shut down! Readers who are about to buy equipment are recommended to "shop around" for the package that suits their requirements best.

3 Preparative HPLC of Phytoalexins

3.1 Principles

Since phytoalexins are inducible rather than constitutive products of plant metabolism, some means of eliciting their synthesis and accumulation must be found. This may take the form of microbial challenge or physicochemical insult. During a period of incubation after the start of the elicitor treatment the concentration of phytoalexin rises, reaches a peak and then often declines. The experimenter wishing to maximise the yields of phytoalexin will generally wait for this peak before extracting. The extraction solvent and the procedures adopted for extraction are of crucial importance in terms of both recovery and complexity of the extract obtained. They therefore reflect on the success and ease of the subsequent HPLC step. Generally some form of clean up of the crude extract is beneficial. The chromatographic separation itself depends on the type of column and the eluting solvent. Fractions are usually cut on the basis of variation in refractive index or UV absorption of the effluent. They must then be tested for antimicrobial activity and purity.

3.1.1 Elicitation

If a compound is classified as a phytoalexin, it must be synthesized in response to microbial challenge, but this is not the preferred method of elicitation when isolating milligram quantities of phytoalexins. Apart from the time-consuming necessity of preparing the challenge inoculum and introducing it to the plant material in an appropriate manner there is the added risk that the parasite will metabolize the phytoalexin to other compounds. Whether or not these are antimicrobial, a false idea of the plant's ability to synthesize phytoalexin will be obtained. A way of avoiding such misleading data is to use abiotic elicitors. There are many chemicals from which to choose which, at appropriate concentrations, will cause phytoalexin elicitation. Of these, the salts of heavy metals (e.g. copper, mercury, and silver) have proved the most popular. Other treatments include the respiratory inhibitors sodium iodoacetate, sodium fluoride, potassium cyanide and 2,4 dinitrophenol, as well as the surfactants Triton X-100, Nonidet P40 and sodium dodecyl sulphate (Bailey 1982). Brief exposure of French bean cotyledons to chloroform vapour has also been reported as successful (Bailey and Berthier 1981). Mechanical damage is usually ineffective although there are exceptions (see Sect. 3.2.1). Local freezing caused phytoalexin accumulation in French bean (Rahe and Arnold 1975) and short wavelength UV irradiation, 254–270 nm, elicited phytoalexins in various legumes as well as grape vine and potato (see Bailey 1982 for review). One feature which all effective means of elicitation seem to have in common is that they cause discrete areas of necrosis (Bailey 1982).

A second area of choice lies in the part of the plant selected for study. Different organs of the same plant may not only accumulate different concentrations of phytoalexins but may synthesize entirely different classes of compound. For example pigeonpea seeds synthesize four isoflavones and the isoflavanone, cajanol (Dahiya et al. 1984), whereas leaves accumulate a chalcone and two isomeric stilbenes (Cooksey et al. 1982). Reciprocally, cotyledons of groundnuts synthesize stilbenes (Aguamah et al. 1981) but leaves accumulate the pterocarpan, medicarpin (Strange et al. 1985).

A third factor to be considered is the time and conditions of incubation after initiation of the elicitation treatment. Generally around 3 days at 20°–25 °C is appropriate for obtaining maximum yields, but much lower concentrations may accumulate at higher or lower temperature. Aeration is important since low oxygen tensions can inhibit or even abolish phytoalexin accumulation (Wotton and Strange 1985).

In order to help the experimenter choose an appropriate mode of elicitation, the type of plant material and optimum conditions for elicitation, some form of preliminary assay must be done. One convenient technique is to use a modification of the TLC assay of Homans and Fuchs (1970). Plant material which has been treated to elicit phytoalexins is extracted and the extracts spotted on a silica gel TLC plate. The plate is developed in a suitable solvent and, after evaporation of the solvent, is sprayed with a spore suspension (ca. 10^7 ml^{-1}) of *Cladosporium cucumerinum* or *C. herbarum* suspended in a liquid medium such as half-strength Czapek Dox. After incubation at high humidity for 48–72 h antifungal compounds show as white zones on a dark background. The size of the white zone is related to the amount of phytoalexin and may therefore be used as a guide to the conditions of elicitation and incubation that are most appropriate, as well as the type of plant material which may be of interest.

3.1.2 Extraction and Clean Up

The aim of extraction is to achieve complete recovery of phytoalexin in solution with as little contamination from extraneous solutes as possible. With some non-chlorophyllous material direction extraction into a non-polar solvent gives excellent results (see Sect. 3.2.2) but with green material this is not feasible as it usually results in an extract heavily contaminated with chlorophyll and accessory photosynthetic pigments. One way to avoid this is to place droplets of elicitor on the surface of green plant material, with or without a wound, and to recover the phytoalexins from the drops without attempting to extract the underlying material. However, this is tedious and inefficient since phytoalexins may accumulate predominantly within the plant tissues rather than the so-called diffusion drops. A more convenient and more efficient method is to vacuum-infiltrate a solvent into the intact material and agitate gently for several hours (Keen 1978). The solvent should be polar enough not to extract the photosynthetic pigments, but not so polar that extraction of the phytoalexins is incomplete. The technique has also been used successfully in the author's laboratory with fleshy material such as cotyledons of soybean, groundnut and peas. Homogenization is quicker but may result in the extraction of greater amounts of undesirable compounds. Homogenates require filtering or centrifuging before progressing with the next step.

Phytoalexins are usually obtained as minor components in aqueous alcoholic extracts of plant material. A method of concentrating the phytoalexins and eliminating as much of the impurities as possible is therefore required. Solvent partitioning is a convenient technique. After removal of the alcohol by film evaporation, the remaining solution is partitioned against a non-polar solvent such as ethylacetate, dichloromethane or ether. The non-polar fraction containing the phytoalexins is usually dried over anhydrous sodium sulphate, evaporated to dryness and dissolved in a suitable volume of solvent compatible with the subsequent HPLC step.

Another technique which has been used for phytoalexin purification prior to isolation on HPLC is open-column chromatography on Sephadex LH20 (Coxon et al. 1980 and see Sect. 3.2.2; Bhattacharyya and Ward 1985 and see Sect. 3.2.3).

Solid phase extraction seems to have been used in connection with phytoalexins mainly for analytical samples (Moesta and Grisebach 1982 and see Sect. 4.2.2; Lyon 1984 and see Sect. 4.2.3; Lee et al. 1986 and see Sect. 4.2.5). However, recent work with toxins from phytopathogenic fungi has shown how powerful this technique can be when applied to concentration and partial purification of small components from large samples (Macko et al. 1985; Clouse et al. 1985).

A further procedure which would seem to merit attention is flash chromatography (Still et al. 1978). Here samples are loaded onto columns containing silica gel (230–400 mesh) and eluted with a suitable solvent under low air pressure. Separation is so rapid (10–20 min) that fractions are normally collected manually. Under favourable conditions individual fractions may contain essentially pure phytoalexin as demonstrated by analytical HPLC (unpublished observations from author's laboratory).

3.1.3 Chromatography

Few investigators have used large-scale preparative HPLC for phytoalexin purification. Generally a semi-preparative column (ca. 25 × 1.0 cm i.d.) is sufficient to obtain a few milligrams for chemical analysis but a larger column (ca. 30 × 6.0 cm i.d.) is useful if greater quantities are required, e.g. for toxicological studies. The choice of packing material is a moot one. Silicas are much cheaper than bonded silicas, but their disadvantage is that if gradient elution has to be used they take a long time to re-equilibrate. Reverse phase columns, e.g. ODS (octadecylsilane functionality), do not have this disadvantage but for some solute mixtures resolution is not as good as the normal phase silicas. The choice of solvent is usually determined by trial runs. Non-polar solvents are used to elute silica columns (see for example Sect. 3.2.3) and polar ones to elute reverse phase columns (see for example Sect. 3.2.2). For large-scale preparative HPLC, solute detection by a refractometer is convenient, since it is not as easily swamped as a UV detector. Fractions may be collected manually according to peaks and troughs on the recorder or alternatively by means of one of the "intelligent" fraction collectors on the market. Fractions are evaporated under reduced pressure and tested for antimicrobial activity, usually by means of the TLC assay (see Sect. 3.1.1). A further run of the active material using the same or different mobile and stationary phases usually results in a sample of sufficient purity for analysis by mass spectrometry and nuclear magnetic resonance spectroscopy.

3.2 Practice

3.2.1 Isolation of Stilbene Phytoalexins from Kernels (After Aguamah et al. 1981)

Groundnuts were shelled and the kernels soaked in water for 7 h. After removing the testa the cotyledons were separated and sterilized in H_2O_2 (30% w/v) for 30 min. The cotyledons were rinsed in sterile distilled water and allowed to imbibe sterile distilled water for 16 h. Phytoalexins were elicited by cutting the kernels into 2-mm thick slices with a sterile multiblade knife. The slices were placed in shallow layers in Petri dishes and incubated in the dark at 25 °C. During the incubation period they were kept well aerated by periodically opening the dishes in a sterile hood and agitating the contents. After 72 h the slices were homogenized in 95% ethanol and the product filtered through Whatman no. 1 filter paper on a Buchner funnel. From this point onwards care was taken to protect the preparation from light as far as possible, as the phytoalexins are light sensitive (Aguamah et al. 1981). The alcohol was removed by film evaporation at <40 °C and the aqueous preparation transferred to a separating funnel. After partitioning three times against equal volumes of petroleum ether (40°–60 °C b.p.) to remove fats, the remaining aqueous phase was partitioned three times against equal volumes of ethylacetate. In the event of the immiscible phases not separating quickly a few ml of saturated brine was added to hasten the process. The combined ethylacetate phases containing the phytoalexins were dried over anhydrous Na_2SO_4 and filtered before HPLC.

Samples were chromatographed on a silica column (30 × 6 cm i.d.) using a Prep LC system 500 (Waters Associates) with petroleum ether (40°–60 °C b.p.)-ethylacetate (7:3) as mobile phase. Fractions were cut in the effluent according

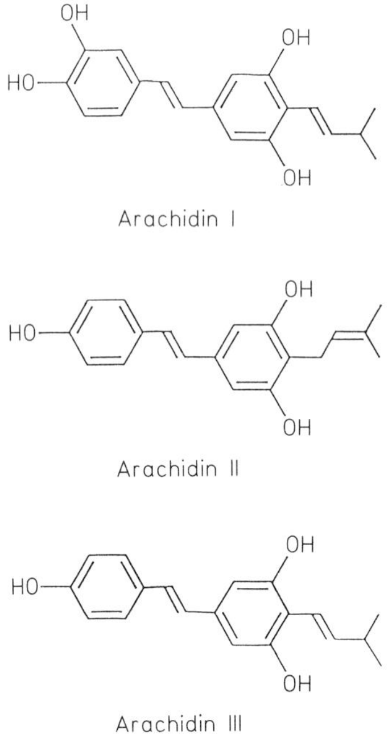

Fig. 3. The arachidins, three stilbene phytoalexins from groundnut kernels

to peaks detected by means of a refractometer. They were evaporated to a small volume and tested for activity using the TLC-*Cladosporium* assay (see Sect. 3.1.1). Solvent was removed from active fractions by film evaporation at <40° and the samples dissolved in acetonitrile. They were further purified by reverse phase HPLC on a column of Hypersil 10 ODS (25 × 10 cm i.d.) protected by a guard column (5.0 × 0.46 cm i.d.) containing Lichroprep RP 18. The mobile phase was acetonitrile-water (1 : 1) and the effluent was monitored at 335 nm. The three stilbene phytoalexins, arachidins I, II, and III (Fig. 3) eluted at 8.0, 9.5, and 10.5 min respectively.

3.2.2 Isolation of Hydroxyflavan Phytoalexins from Narcissus Bulbs (Coxon et al. 1980; O'Neill and Mansfield 1982)

Fleshy scales of daffodil bulbs (cv. Golden Harvest) were excised and the adaxial epidermis removed. The scales were placed adaxial surface uppermost on moist tissue paper in plastic sandwich boxes. Phytoalexins were elicited biotically by pipetting droplets (20 µl) of an aqueous spore suspension (10^5 spores ml^{-1}) of *Botrytis cinerea* on the stripped surfaces of the scales and incubating for 5 days in the closed boxes at 18 °C in the dark. Tissue was scraped from the inoculum sites with a scalpel and stored at −20 °C. Phytoalexins were extracted from the infected tissue by homogenizing in redistilled ether (at least 10 ml ether g^{-1} fresh weight) and left to soak for 3 h at 4 °C. The supernatant was retained and the homogenate washed twice with ether. Ether extracts were combined and evapo-

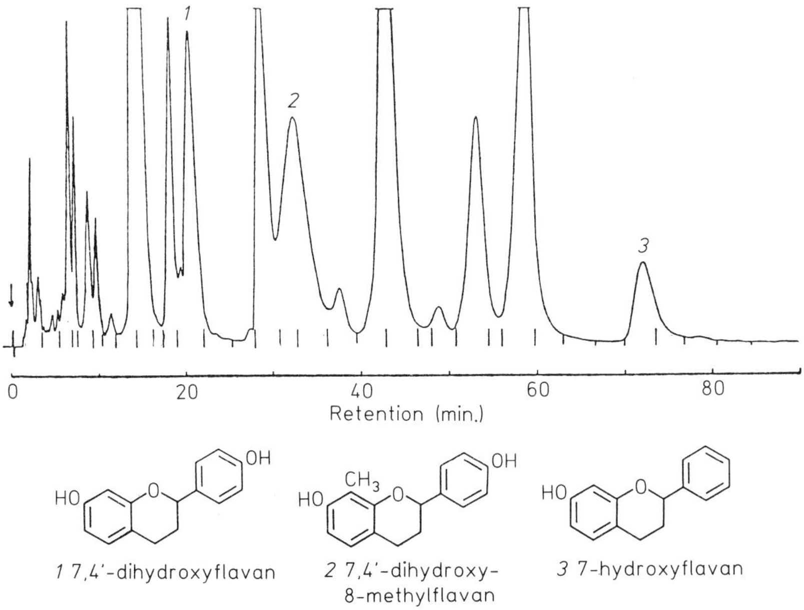

Fig. 4. Separation of hydroxyflavan phytoalexins from infected narcissus bulbs by reverse phase HPLC. For details see text

rated at 30 °C. In early experiments the ether extracts were subjected to a clean up by open column chromatography on a column of Sephadex LH 20 (70 × 2.5 cm) with methanol as mobile phase and by preparative TLC on silica gel plates. Subsequently a one step HPLC procedure was used. The ether extract of 40 g of scales was evaporated to dryness and taken up in 1 ml methanol. Aliquots (25 µl) were applied to a column of Hypersil ODS 5 µm (20 × 0.8 cm i.d.) and eluted with a solvent consisting of 35% methanol in 5% formic acid at 5 ml min^{-1}. The eluate was monitored at $\lambda = 284$ nm (Fig. 4). 7,4′-dihydroxyflavan, 7,4′-dihydroxy-8-methylflavan and 7-hydroxyflavan eluted at ca. 20, 33, and 72 min respectively. Other antifungal compounds were present but were not characterised further.

3.2.3 Isolation of Pterocarpan Phytoalexins (Glyceollins) from Soybean Cotyledons (After Epperlein et al. 1986; Battacharyya and Ward 1985)

Soybeans seeds (ca. 50 g) were surface-sterilized by immersion in H_2O_2 (10% w/v) for 30 min. They were rinsed thoroughly in sterile distilled water and allowed to imbibe sterile distilled water for 16 h at 25 °C. Testas were removed and the separated cotyledons placed concave side down in a single layer on the bottom of 9-cm Petri dishes (3–5 g per dish). They were incubated with $AgNO_3$ solution (3 ml; 10 mM) in the dark at 25 °C for 72 h and extracted in 95% ethanol by the facilitated diffusion technique of Keen (1978) as follows: cotyledons were immersed in 95% ethanol and thoroughly infiltrated with the solvent by alternating the samples between low and atmospheric pressure. This may be done conveniently in a Buchner flask. The top of the flask should contain an air-tight bung with a central hole occupied by a short length of glass tubing. A piece of flexible tubing was fitted to the top of the glass tubing and closed with a spring clip. The side arm of the flask was attached to a vacuum pump (e.g. water pump) and a vacuum applied. When no more bubbles rose from the sample, but before the solvent began to boil, the vacuum was gently released by means of the spring clip. The process was repeated several times until the material was thorougly infiltrated. After infiltrating the soybean cotyledons by this technique they were incubated at 4 °C for 16 h. The solvent was then decanted and the alcohol removed by film evaporation at <40 °C. The remaining aqueous solution was made up to ca. 50 ml with water and partitioned three times against equal volumes of ethylacetate, a few ml of saturated brine being added if the phases did not separate easily. The ethyl acetate phases were combined, dried over anhydrous Na_2SO_4 and evaporated to give a volume equivalent to 2 g cotyledons ml^{-1}. Samples (0.5–1.0 ml) were injected into an HPLC instrument consisting of an Altex sample injection valve, a precolumn (5.0 × 0.46 cm i.d.) containing spherisorb SGP and a main column (25.0 × 1.0 cm i.d.) packed with Partisil 5. Samples were eluted in hexane-ethyl acetate (4:1) at a flow rate of 4 ml min^{-1}. The effluent was monitored at 290 nm with a Pye-Unicam UV detector and the absorbance plotted on a Tekman Pontiometric Chart recorder. Glyceollins I, II, and III eluted at ca. 23.8, 25.8, and 29.6 min respectively (Fig. 5).

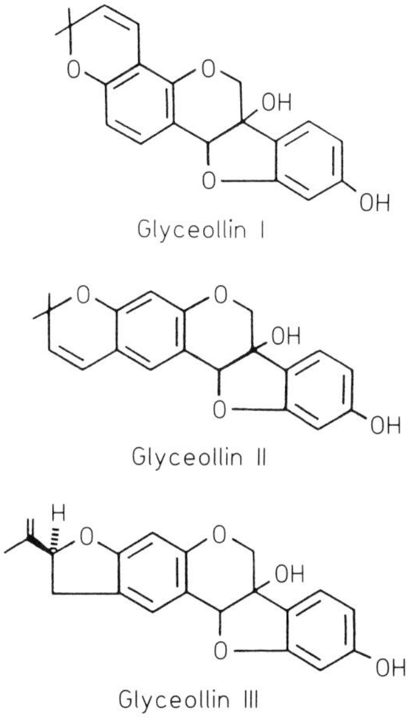

Fig. 5. The glyceollins, three isomeric pterocarpan phytoalexins from soybean

A rather different protocol has been adopted by Bhattacharyya and Ward (1985). Soybean seedlings were grown for 8 days under a 16-h photoperiod (ca. 300 μE $m^{-1}s^{-1}$) at a maximum temperature of 22.5 °C in the light and a minimum of 16 °C in the dark. Phytoalexins were elicited biotically by applying 25 μl of a zoospore suspension (10^5 ml^{-1}) of *Phytophthora megasperma* f.sp. *glycinea* race 1 to the wounded dorsal side of the detached cotyledons and incubating for 48 h at 25 °C and 100% RH in the dark. The glyceollins were extracted with 95% ethanol for 10 min at 100 °C and the extract was filtered through cheesecloth. The filtrate was evaporated to near dryness under reduced pressure and the residue dissolved in water. Glyceollins were extracted by partitioning the aqueous solution three times with equal volumes of ethyl acetate. The ethyl acetate fraction was concentrated and the glyceollin isomers were partially separated on a column of Sephadex LH20 (86 × 2.4 cm i.d.; void volume 138 ml) with methanol as eluant. The elution volumes of the isomers which were recognised by their UV spectra were glyceollin I, 517 ml; glyceollin II 538 ml; glyceollin III, 559 ml. Fractions were combined accordingly and purified by HPLC on a semi-preparative silica column (Whatman Partisil 10 M 9/50 50 cm × 0.94 cm i.d.) with 3% or 5% isopropanol in hexane as the mobile phase. Effluent was monitored at 285 nm and the elution times for glyceollins I, II, and III respectively were 52.1–55.8, 59.6–60.9, 65.6–67.0 min with 3% isopropanol (flow rate 5 ml min^{-1}) and 35.3–39.3, 38.8–42.7, 43.5–46.3 min with 5% isopropanol (flow rate 3 ml min^{-1}). Peaks corresponding to the individual gylceollins were cut and recycled four to six times before collecting in order to ensure purity.

3.2.4 Isolation of Isoflavone and Isoflavanone Phytoalexins from Pigeonpea Seeds (After Dahiya et al. 1984)

Seeds of pigeonpea (*Cajanus cajan* c.v. Prabhat: 50 g) were soaked in distilled water at 25 °C for 24 h and cut into slices ca. 2 mm thick with a sharp scalpel. The sliced seeds were moistened with 0.01 M potassium phosphate buffer (pH 7.5) and incubated under non-sterile conditions at 25 °C for 4 days in order to allow a microflora to develop and elicit the phytoalexins. Samples were homogenized in 80% methanol (1.5 w/v) and the homogenate centrifuged at 10,000 *g* for 30 min. The supernatant was decanted and the residue suspended in 80% methanol before centrifuging once more. Methanol was removed from the combined supernatants by film evaporation at <40 °C. The aqueous solution was partitioned three times against equal volumes of petroleum ether (40°–60 °C b.p.; 50 ml) to remove fats, and three times against equal volumes of ethyl acetate. The aqueous and petroleum ether fractions were discarded and the combined ethyl acetate fractions dried over anhydrous Na_2SO_4. After concentration by film evaporation, the ethyl acetate fraction was separated by HPLC. The stationary phase was a column of Hypersil 10 ODS (25 × 1.0 cm i.d.) which was protected with a guard column (5 × 0.46 cm i.d.) packed with Lichroprep RP 18 (24–40 µm particle size). Samples (up to 2 ml) were injected by means of an Altex injection

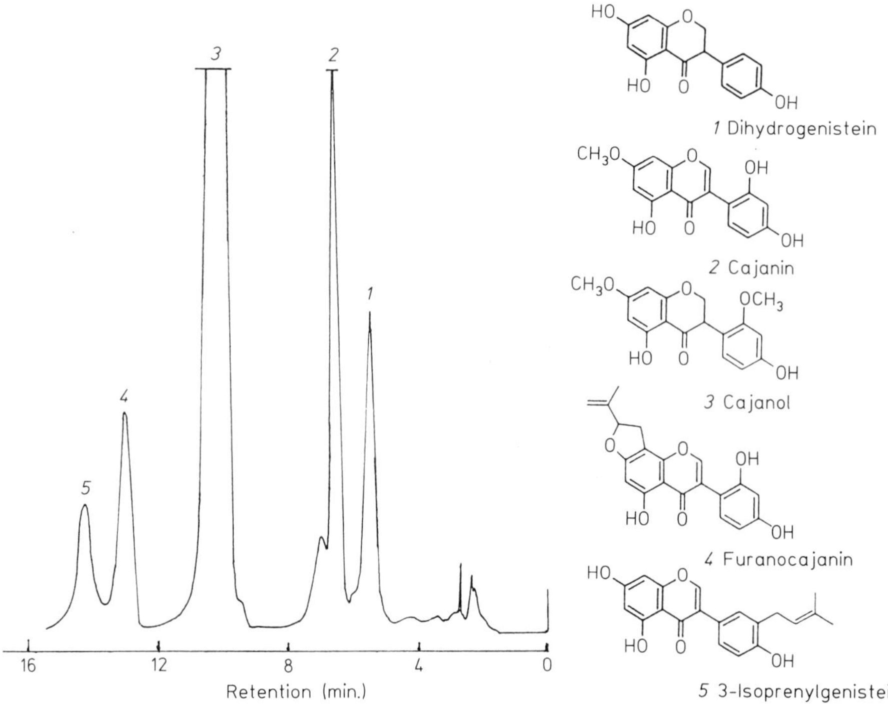

Fig. 6. Separation of four isoflavone and an isoflavanone phytoalexin from seeds of pigeonpea (*Cajanus cajan*). For details see text

valve and eluted in acetonitrile : water (1 : 1; v/v) at a flow rate of 4 ml min^{-1}. The effluent was monitored by a Pye-Unicam LC UV detector set at 290 nm. Phytoalexins eluted after 5.6, 6.7, 10.2, 12.9, and 14.2 min (Fig. 6).

3.2.5 Isolation of Chalcone and Stilbene Phytoalexins from Pigeonpea Leaves (Cooksey et al. 1982, 1983)

Leaves from pigeonpea plants (*Cajanus cajan* c.v. Prabhat) grown for 5–6 weeks in the greenhouse were detached and surface sterilised in H_2O_2 (7.5% w/v) for 15–20 min. They were rinsed in three changes of sterile distilled water and the excess water removed by absorption on sterile muslin. Wounds (2–3 mm long; 20 wounds/leaf) were made in the intercostal areas of the leaves and spore suspension of *Botrytis cinerea* (5.6×10^5 spores ml^{-1}; 20 μl per wound) placed on each wound. The inoculated leaves were incubated at high humidity for 48–72 h in the dark. Alternatively, phytoalexins were elicited by substituting sucrose (100 μg ml^{-1}) for the spore suspension in the 20 μl droplets. Phytoalexins were extracted from leaves by the facilitated diffusion technique (Keen 1978; and see Sect. 3.2.3) in 95% ethanol. The ethanol was evaporated and the residue dissolved in acetonitrile-water-acetic acid (65 : 35 : 1; v/v). Samples (1 ml) were injected into an HPLC instrument as described in the previous section (3.2.4) and eluted with the

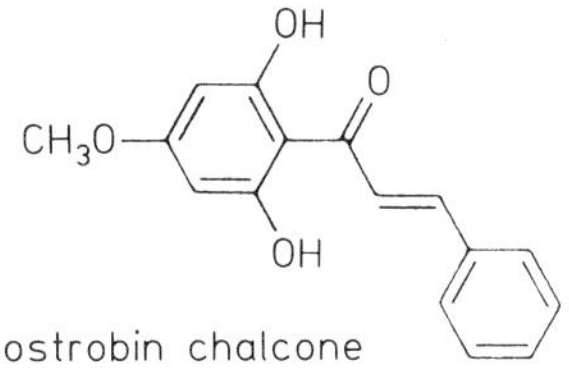

Pinostrobin chalcone

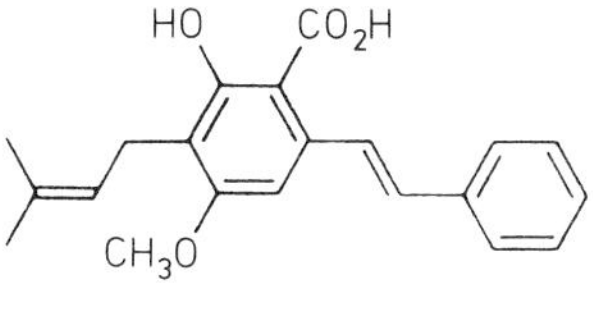

2-carboxy-3-hydroxy-4-isoprenyl-5-methoxystilbene

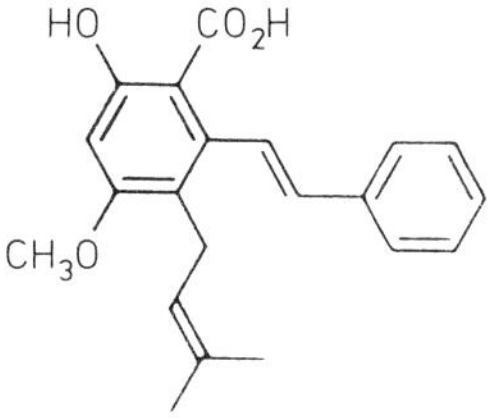

2-carboxy-3-hydroxy-6-isoprenyl-5-methoxystilbene

Fig. 7. Pinostrobin chalcone and two isomeric stilbene-2-carboxylic acid phytoalexins from leaves of pigeonpea (*Cajanus cajan*)

same solvent as used for dissolving the sample. The effluent was monitored at 310 nm. Pinostrobin chalcone and the two isomeric stilbene carboxylic acids eluted after 7.5, 14.1, and 16.8 min respectively (Fig. 7).

4 Analytical HPLC of Phytoalexins

4.1 Principles

It is in the analytical field that HPLC techniques can perhaps make the greatest contribution to phytoalexin studies. The comparative ease of sample preparation combined with automated methodologies bring into range the feasibility of extensive studies. These include determination of the potential of various genotypes of the plant to synthesize phytoalexins, the effect of environmental conditions on this potential, the rate of phytoalexin accumulation on challenge with microorganisms or abiotic elicitors and the rate at which they may be metabolised by host or parasite.

4.1.1 Elicitation

Elicitation of phytoalexins by abiotic means (see Sect. 3.1.1) is most appropriate for a comparison amongst cultivars of the potential to synthesize and accumulate phytoalexins. Such comparisons should emphasise rates of phytoalexin accumulation as an early response to elicitation, as these are more important in determining the outcome of microbial challenge than the absolute amounts accumulated over a longer period (Mansfield 1982; Wotton and Strange 1985). It may also be important to compare cultivars for phytoalexin accumulation under varying degrees of stress, either before elicitation (e.g. drought) or during elicitation and the subsequent period of incubation (e.g. low O_2 tension). Microbial challenge can then be used as the method of elicitation and the results compared with those obtained for potential accumulation. If the concentration elicited by a microorganism is lower than the potential, the possibilities arise that the microorganisms may be inhibiting phytoalexin synthesis or degrading the phytoalexin as soon as it is formed.

4.1.2 Sampling

The most important point that an experimenter will wish to establish is the cellular concentration of the phytoalexin. In particular, in infection studies, he will need to know what this concentration is in the vicinity of the infecting microorganism. Only then will he be able to answer the question whether or not the phytoalexin has reached a sufficient concentration to inhibit its growth. This represents a formidable problem in sampling. In laminar tissues inoculated with aqueous droplets containing the challenge inoculum, the droplets themselves may be harvested separately and analysed. The underlying tissue may then be separated from healthy tissue by punching it out with a cork borer. With thicker tissue some means of separating infected from uninfected layers of tissue must be found. Yoshikawa et al. (1978), for example, froze hypocotyls of soybean inoculated with *Phytophthora megasperma* var *sojae* (= *Phytophthora megasperma* f.sp. *glycinea*) and cut sections 0.25 mm thick with a freezing microtome which were analysed separately. Although these techniques serve to minimize the amount of non-infected tissue harvested with the infected tissue, there is no satisfactory method for eliminating it altogether. Thus, phytoalexin concentrations recorded from extracted material tend to represent minimum rather than maximum values.

Monitoring the growth of the parasite in the plant lies outside the scope of this chapter and the reader is referred to the microscopy techniques of Rohringer et al. (1977) and the chemical assays of Ride and Drysdale (1972) and Seitz et al. (1979). It need hardly be stressed that these studies should go hand in hand with those of phytoalexin accumulation in response to microbial challenge.

Another problem is the reference weight of the sample. Commonly, phytoalexin yields are presented in terms of μg phytoalexin g^{-1} fresh weight of sample but with small amounts of material which may be wet with the residues of inoculation droplets or which may be "mushy" owing to degradation by the parasite this is subject to error. One method is to freeze-dry the material and express yields in terms of the freeze-dried weight of the sample (Lyon 1984). However, the more important parameter to establish is the volume of material containing the phytoalexin. Phytoalexin concentration may then be recorded in μg ml^{-1}. In practice this seldom gives results greatly at variance with the more conventional μg g^{-1} fresh weight, since most plant tissue has a specific gravity of about 1.

4.1.3 Extraction and Clean Up

When extracting phytoalexins for quantitative assay, the main considerations are efficiency, selectivity and recovery. Although homogenization is frequently employed and is generally very efficient, it is cumbersome if large numbers of samples have to be processed. It can also cause the release of unwanted components into solution which may interfere with the subsequent HPLC. Some workers have therefore preferred to use techniques in which the tissue is kept intact such as the facilitated diffusion procedure of Keen (1978; and see Sect. 3.2.3) or simply boiling the tissue in the extracting solvent (Moesta and Grisebach 1982; Goossens and Van Laere 1983; and see Sects. 4.2.2 and 4.2.4). In all cases the choice of extracting solvent is crucial in determining the type of solute extracted, i.e. selectivity. Generally, workers in the phytoalexin field have chosen rather polar solvents such as aqueous alcohols. After removal of the alcohol by evaporation, the phytoalexins may be partitioned into a less polar solvent such as ethyl acetate. Normally this is done in a separating funnel but a new device termed a Mixxor (Cole-Parmer Instrument Company, 7425 North Oak Park Avenue, Chicago, Illinois 60648) offers a more convenient method for achieving the same end. Vessels of differing capacity may be obtained ranging from 2–20 ml. Alternatively, phytoalexins may be extracted directly into a relatively non-polar solvent. For example Coxon et al. (1980) and O'Neill and Mansfield (1982) have used ether in their work on phytoalexins from narcissus bulbs.

It is difficult to know how much material is lost during extraction. One solution is to divide batches of similar samples into two and add a known amount of pure phytoalexin to one and not to the other. A correction factor by which results from samples must be multiplied may then be calculated as follows:

$$x = \frac{ax - b}{c}$$

where x = correction factor
a = amount of phytoalexin found in samples with phytoalexin addition
b = amount of phytoalexin added to samples
c = amount of phytoalexin found in samples without phytoalexin addition.

An alternative is to add a compound of similar structure to the phytoalexins which can serve not only to derive a correction factor for losses during extraction but also as an internal standard in the subsequent HPLC (Goossens and Van Laere 1983; and see Sect. 4.2.4).

One method of cleaning up samples prior to HPLC is to use Sep Pak cartridges (Waters Associates). These are available with high polarity (silica or florisil) or low polarity (C_{18} bonded silica) packing material. Samples may be concentrated on the cartridges and selectively eluted ready for HPLC analysis (see Sects. 4.2.2, 4.2.3, and 4.2.5).

4.1.4 Chromatography

Generally the experimenter interested in analytical HPLC of phytoalexins will have large numbers of samples to analyse. An autosampler is therefore essential. For the same reason he will wish to reduce analysis time as far as possible. Usually this means that he will have to employ gradient elution. Several authors (Goossens and Van Laere 1983; Lyon 1984; and see Sects. 4.2.3 and 4.2.4) have used gradient elution of normal phase columns as they give high resolution. Their disadvantage is that they generally require a long time to re-equilibrate. Goossens and Van Laere (1983) have overcome this problem by using isohydric methanol in their elution gradient. This allowed the water content of the silica column to remain constant during elution. Re-equilibration time was reduced to 5 min and peak tailing, which occurred when dry methanol was used, was eliminated. Most other workers have used ODS columns which require only minimal re-equilibration time. Whichever type of column is decided upon it is advisable to protect the main column with a short guard column unless a vigorous clean up procedure, e.g. Sep Pak has been used.

The choice of solvent as eluant is of key importance. Mansfield et al. (1980) reported the separation of seven furanoacetylenic phytoalexins of *Vicia faba* using an ODS column and a solvent gradient. The initial conditions were methanol-5% formic acid (7:13) running linearly over 30 min to methanol-acetonitrile-5% formic acid (4:1:5). Using this system, the last phytoalexin peak eluted at ca. 45 min and the internal standard (n-butylsalicylate) at ca. 60 min. Subsequently an isocratic system (acetonitrile – 1% formic acid 60:40) was used which, in conjunction with a smaller volume of stationary phase, cut this time by more than a third (Mansfield et al. 1984). However, large differences in the polarity of components of a phytoalexin mixture may make isocratic elution impossible, e.g. phaseollin and kievitone from *Phaseolus vulgaris* (Goossens, personal communication). A further constraint is placed on solvent composition if short wavelength UV (e.g. 210 nm) is used for detection, since many solvents have appreciable absorption at this wavelength.

Correct identification of peaks is mandatory. Many experimenters rely upon retention time and UV absorption at an appropriate wavelength, but Porter et al. (1979) found this insufficient for the *Vicia faba* phytoalexins. In their gradient elution system, even with thermostatic control of eluent and column temperatures, the 95% confidence ranges of the elution times of wyerone epoxide and wyerone acid overlapped as did dihydrowyerone acid and wyerol acid. Peak identities had to be confirmed therefore by spiking with authentic phytoalexins. For increased confidence of peak identification and purity experimenters may wish to invest in one of the more sophisticated detection devices described in Sect. 2.

4.2 Practice

4.2.1 Analytical HPLC of Furanoacetylenic Phytoalexins from Broad Bean (Methods of Porter et al. 1979; Mansfield et al. 1980, 1984)

Phytoalexins were elicited with spore suspension of *Botrytis cinerea* or *B. fabae* (ca. 2000 and 40 conidia/20 μl droplet, respectively) and $HgCl_2$ 10^{-2} M (20 μl droplets). Fungal inoculum was applied to imbibed cotyledons and detached field-grown leaves and pods and $HgCl_2$ to imbibed cotyledons. After incubation at high humidity for various times, inoculum droplets were collected with a Pasteur pipette. Leaf tissue underlying the inoculum droplets was cut out with a cork borer. Cotyledon tissue not more than 1 mm thick was sliced off with a razor blade and pod endocarp tissue was removed by scraping with a spatula to a depth

ble 1. Calibration equations for the Wyerone phytoalexins. (Porter et al. 1979). $y=ax+b$, where y is amount (μg of phytoalexin injected and x is the ratio of the phytoalexin peak area (330 nm) to that m 30 μg n-butyl salicylate

ytoalexin	a (slope)	b (intercept)	Correlation coefficient	S.D. slope[a]	S.D. intercept[a]	S.D. for 1 μg	95% confidence limits for 1 μg (μg)
yerone epoxide (1)	2.7276	−0.00883	0.9998	0.00698	0.00307	0.0033	±0.01
yerone acid (2)	3.6235	+0.01484	0.9995	0.02947	0.00980	0.0106	±0.02
hydrowyerone acid (3)	1.6931	−0.01094	0.9913	0.02788	0.00499	0.0097	±0.02
yerol (4)	3.0795	−0.01401	0.9983	0.02513	0.00767	0.0084	±0.02
hydrowyerol (5)	2.3964	−0.01559	0.9975	0.02243	0.00290	0.0056	±0.01
yerone (6)	2.7328	+0.03313	0.9997	0.00724	0.00634	0.0064	±0.01
hydrowyerone (7)	1.9720	−0.00167	0.9997	0.00592	0.00358	0.0069	±0.01

$y=\sqrt{x^2\sigma^2 \text{ slope}+\sigma^2 \text{ intercept}}$.

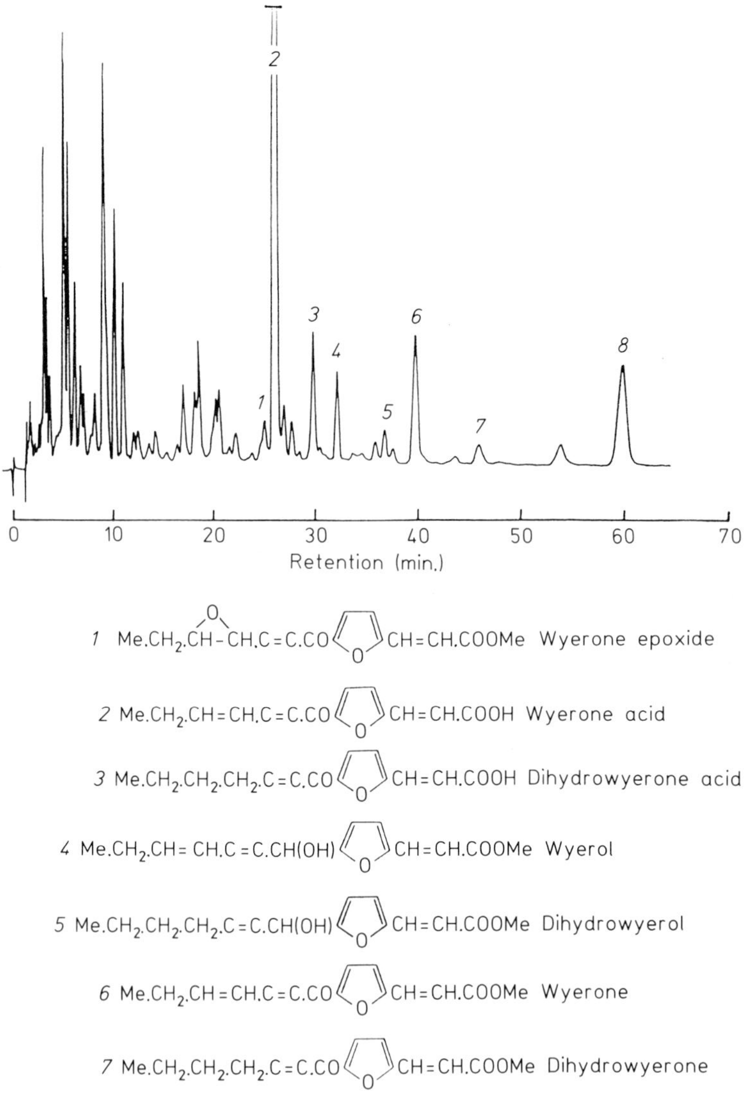

Fig. 8. Separation of furanoacetylenic phytoalexins of *Vicia faba* by gradient elution of a reverse phase column. Peak *8* is the internal standard n-butylsalicylate. For details see text

of about 3 mm beneath inoculum droplets. Tissue and droplets were stored at −20 °C before extraction. Droplets were partitioned three times against twice their volume of ether, and tissue was homogenized in ether. The homogenate was centrifuged and the pellet washed three times with ether. After combining, the ether fractions were taken to dryness in vacuo and the residues dissolved in methanol (1 ml methanol per 0.25–8 g original tissue) containing 3 mg ml^{-1} n-butylsalicylate as an internal standard. Aliquots (10 µl) were injected onto an ODS Hy-

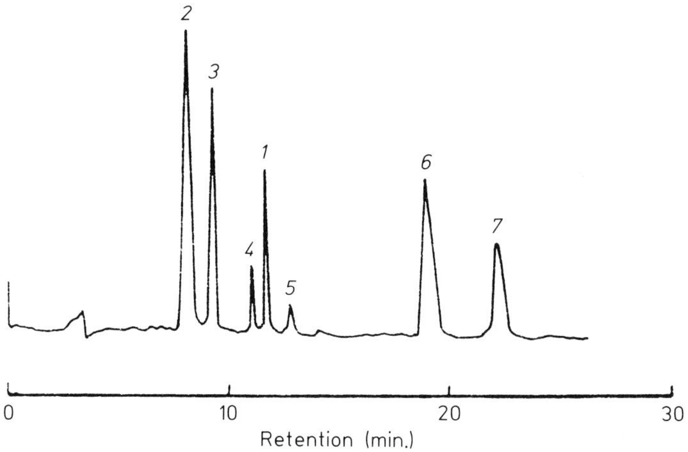

Fig. 9. Separation of furanoacetylenic phytoalexins of *Vicia faba* by isocratic elution of a reverse phase column. For details see text. For identities of peaks see Fig. 8

persil (5 μm) column (20 × 0.8 cm i.d.) fitted with an on-column needle through septum injector. Gradient elution was carried out at 30 °C with a flow rate of 5 ml min^{-1}. The initial solvent conditions were methanol-5% formic acid (7 : 13) running linearly over 30 min to final conditions of methanol-acetonitrile-5% formic acid (4 : 1 : 5). Peaks corresponding to the phytoalexin were detected by UV absorbance at 330 nm (0.1 A units for full-scale deflection). Peak areas were measured by a computing integrator and the phytoalexin amounts calculated from calibration equation derived from peak areas of standards expressed as ratios of the peak area of the butylsalicylate internal standard (Table 1, Fig. 8).

More recently, Mansfield et al. (1984) have described a simpler isocratic system. In this instance an ODS ultrasphere column (25 × 0.46 cm i.d.) was used with acetonitrile – 1% formic acid (60 : 40) as mobile phase. With a flow rate of 0.8 ml min^{-1} the last phytoalexin dihydrowyerone, eluted at ca. 22 min and the internal standard n-butylsalicylate at 38.7 min (Fig. 9).

4.2.2 Analytical HPLC of Pterocarpan Phytoalexins (Glyceollins) from Soybean (After Moesta and Grisebach 1982; Epperlein et al. 1986; Osswald 1985)

Hypocotyls of seedlings of *Glycine max* cv. Harosoy 63 were wounded and inoculated with mycelium of race 1 or race 3 of *Phytophthora megasperma* f.sp. *glycinea*. The seedlings were incubated in the light at 100% humidity for 24 h. Infected segments with 5 mm of healthy material on either side from ten hypocotyls were boiled in a test tube with ethanol (5 ml) for 2 min. The ethanol was decanted and the extraction repeated before finally washing the tissue with cold ethanol (5 ml). The extracts were evaporated to dryness and taken up in 2 × 1 ml chloroform. They were applied to Sep-Pak silica cartridges and non-polar materials eluted with chloroform (5 ml). The phytoalexins were eluted with 3 ml chloro-

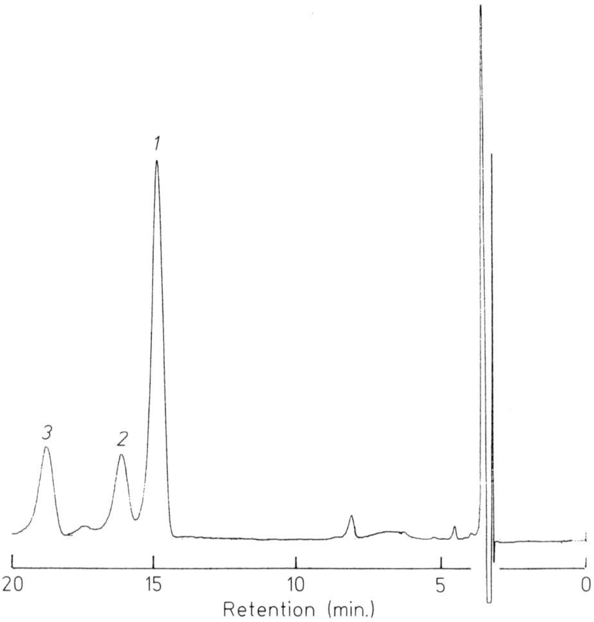

Fig. 10. Separation of glyceollins by normal phase HPLC. For structures see Fig. 5 and for details see text

form-isopropanol (1:1). Samples were evaporated to dryness and the residue taken up in isopropanol (400 µl) before HPLC (see below).

Phytoalexins elicited in soybean cotyledons (3 g samples) by various treatments including $AgNO_3$ were extracted and prepared for HPLC essentially as previously described (see Sect. 3.2.3). Ethyl acetate fractions were dried in vacuo and dissolved in toluene (1.5 ml). Samples were introduced into the HPLC via an Altex sample injection valve fitted with a 20 µl sample loop and chromatographed as previously described except that the main column dimensions were 25 × 0.46 cm i.d. Glyceollins I, II, and III eluted at 14.9, 16.2, and 18.8 min respectively (Fig. 10). The concentration of each isomer was calculated by measuring the peak area (peak height × peak width at half peak height) and comparing it with standards of the pure compounds made up using molar extinction coefficients of 8300 for glyceollin I (Sims et al. 1972), 8700 for glyceollin II, and 9400 for glyceollin III (Lyne et al. 1976). The molar extinction coefficient of glyceollin I has now been revised to the higher figure of 10800 (Bhattacharyya and Ward 1985).

Osswald (1985) has described a reverse phase system for separating the glyceollins, in which the running time was 16 min. The column (75 × 4.6 mm i.d.) contained 3 µm Ultrasphere ODS and the mobile phase consisted of two gradients of A water and B methanol (0–4 min 26%; 4–10 min 26–55% B). Glyceollins I, II, and III eluted at 14.46, 14.77, and 15.20 min at a flow rate of 1.2 ml min^{-1}.

4.2.3 Analytical HPLC of Phytoalexins from Potato (Method of Lyon 1984)

Tubers of *Solanum tuberosum* c.v. Pentland Crown were washed and surface sterilized in 2% Chloros (I.C.I. Ltd. 0.2% w/v available chlorine) for 5 min. After rinsing in water they were wounded with a 4 × 6.4 mm "pop" rivet (Tucker Fasteners Ltd., Birmingham, England) and bacterial suspension added. The tubers were stored in the dark for 5 days at various temperatures in an atmosphere of 10% oxygen in nitrogen. Rots were scraped from tubers and freeze-dried. The dried samples were weighed and stored at −35 °C until analysed. They were ground in methanol (20 ml g^{-1}) centrifuged and the pellet re-extracted with methanol. The two methanol supernatants were combined and evaporated to dryness in vacuo at 40 °C. Samples were dissolved in 2 × 2.5 ml of ethyl acetate and applied to a Sep-Pak silica cartridge (Waters Associates). The phytoalexins were eluted with ethyl acetate (20 ml), the solvent removed in vacuo at 40 °C and the residue transferred in a small volume of ethyl acetate to a glass vial. The sample was dried in a current of air and dissolved in 3% isopropanol in hexane (10 g dry weight of rot ml^{-1}). Samples (20 µl) were injected onto an Ultrasphere silica (5 µm) column (25 × 0.46 mm i.d.) and eluted in a gradient of 3% to 25% isopropanol in hexane over a period of 20 min. The isopropanol hexane mixture was kept at 25% isopropanol for 10 min before re-equilibrating the column with 3% isopropanol in hexane prior to another run. Effluent from the column was moni-

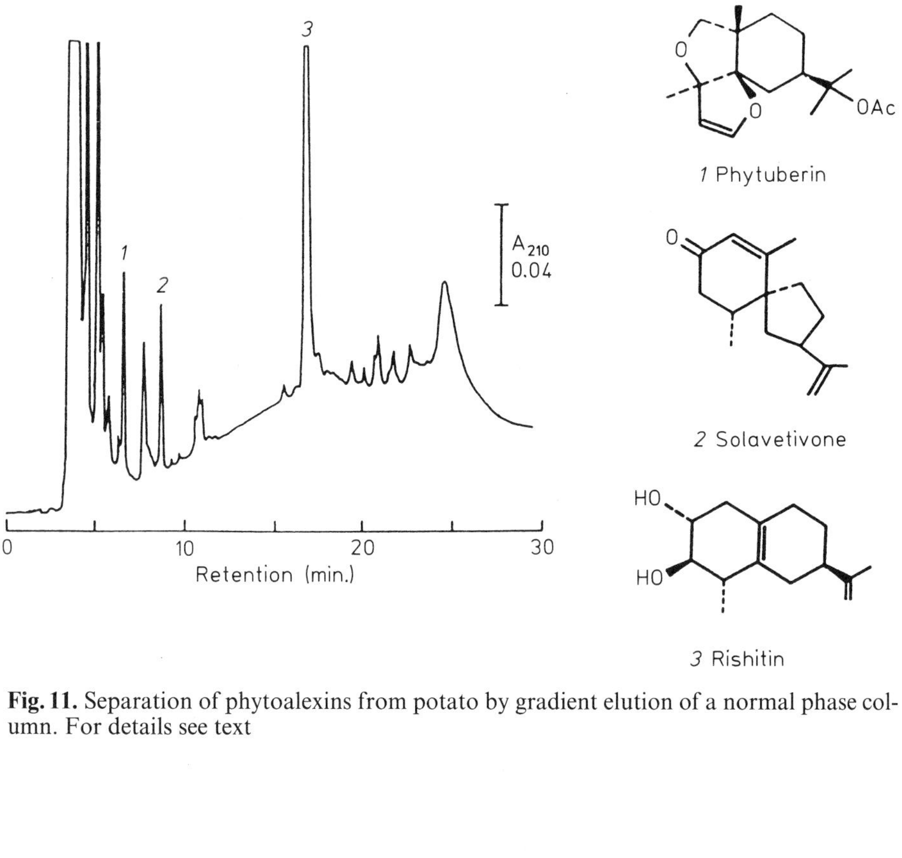

Fig. 11. Separation of phytoalexins from potato by gradient elution of a normal phase column. For details see text

tored at 210 nm and peak areas integrated with a C-R1A integrator. These were compared with areas given by known weights of authentic samples of rishitin, phytuberin, and solavetivone (Fig. 11).

4.2.4 Analytical HPLC of Phytoalexins from French Bean (Method of Goossens and Van Laere 1983; Goossens 1985, personal communication)

Cotyledons, hypocotyls and roots of *Phaseolus vulgaris* etiolated or light-grown were treated in various ways to elicit phytoalexin (e.g. immersion of cotyledons in 10^{-3} M solution of $HgCl_2$) and incubated. The material was frozen in liquid nitrogen and stored at -20 °C. A known weight of vesidryl (2′,4,4′ trimethoxychalcone) was added to the samples, which were then extracted either by boiling or homogenizing in ethanol. Supernatants (obtained after centrifuging if samples were homogenized) were decanted and the extraction process repeated. The combined supernatants were evaporated in vacuo at 40 °C and a suitable volume of water added, depending on sample size. The aqueous fraction was extracted twice with equal volumes of dichloromethane. After removing most of the solvent in vacuo at 40 °C, the sample was transferred to a small vial. The remaining solvent was removed under a stream of N_2 and the sample redissolved in 1 ml chloroform or dichloromethane.

For routine analysis a guard column (6.0 × 0.21 cm i.d.) packed with spherical Perisorb A (35–50 μm) was used to adsorb polar materials. The main column was silica [Lichrospher SI 100 (10 μm) 30 × 0.3 m i.d.]. Samples (25 μl) were eluted at a flow rate of 2 ml min^{-1} with a linear gradient from 1–21 min after injection. The initial composition of the eluant was 100% hexane-chloroform (2:1) and the final composition 100% chloroform-methanol (100:8) (Fig. 12). More recently Goossens (personal communication) has used another silica column (R sil 5 μm) of only 10.0 × 0.46 cm i.d. with a similar gradient. This has allowed the time for analysis and re-equilibration to be cut to 11 min.

4.2.5 Analytical HPLC of Phytoalexins from Cotton (Method of Lee et al. 1986)

Cotton plants (*Gossypium hirsutum* c.v. Acala STC-1 or *Gossypium barbadense* c.v. Seabrook Sea Island) were grown in the greenhouse. Phytoalexins were elicited with a mild strain of *Verticillium dahliae* Kleb. by puncture inoculation with conidial suspensions (2×10^6 ml^{-1}). Inoculated stems were harvested 3–5 days later and the cortical tissue stripped, leaving the stelar tissues free of any glands. Steles were stored at -20 °C until extraction. During extraction procedures precautions were taken to provide protection from direct light, high temperature and basic conditions in order to avoid oxidation. The steles were cut into small pieces and homogenized for 4 min in cold ethylacetate containing 0.1% acetic acid. The homogenate was filtered through a sintered glass funnel and the residue extracted twice more with the same solvent. After pooling, the extracts were evaporated to dryness on a rotary evaporator at room temperature and dissolved in acetonitrile

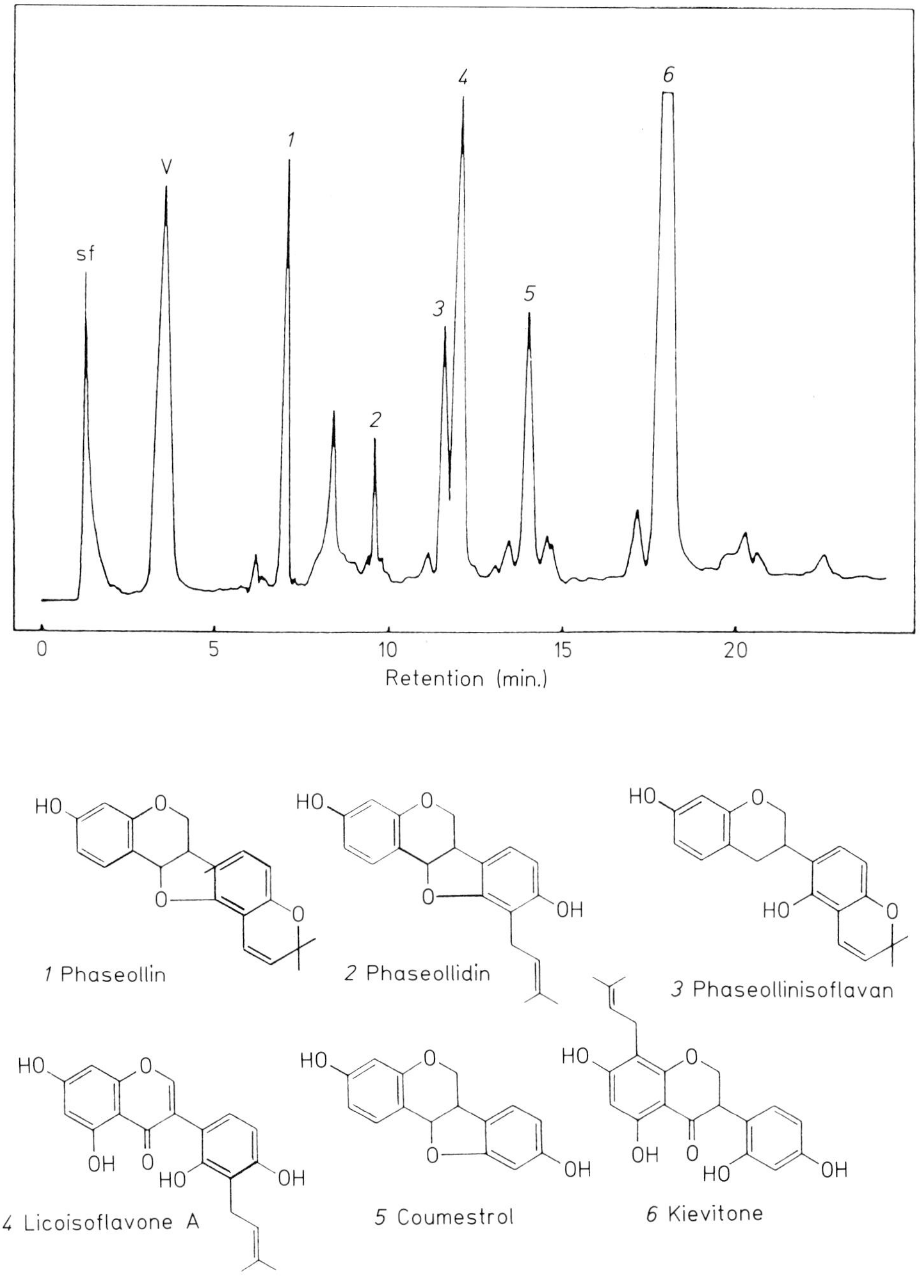

Fig. 12. Separation of phytoalexins from *Phaseolus vulgaris* by gradient elution of a normal phase column. Sf is the solvent front and V is vesidryl, the internal standard. For details see text

with the aid of sonication (2.0 g fresh weight ml^{-1}). Particulate matter was removed with a membrane filter (0.45 μm, Waters).

In one technique 4,4′ bis (N,N-dimethylamino) benzophenone was added to the samples to act as an internal standard. Samples (100 μl) were diluted with distilled water (3 ml) and purified on a Sep Pak C18 cartridge (Waters) which had been pretreated by wetting with 5.0 ml of methanol containing 0.1% acetic acid followed by 3 ml distilled water. The sample was allowed to drip through the column under gravity and non adsorbed material was eluted with 50% methanol containing 0.1% acetic acid. The phytoalexins were eluted in methanol containing 0.1% acetic acid.

Eluate from the Sep Pak (20 μl) was chromatographed on a reverse phase column (Waters Radial PAK C8, 100 × 8.0 mm i.d.). The mobile phase was a concave gradient of 0.1% phosphoric acid-methanol (20–100% methanol over 30 min) delivered according to #8 of Model 660 Waters solvent programmer. The phytoalexins were detected by UV absorption at 254 nm. Under these conditions the internal standard eluted at 29.1 min, desoxyhemigossypol at 3.5 min, hemigossypol at 9.0 min, 6 methoxydesoxyhemigossypol at 11.3 min and desoxyhemigossypol at 13.7 min. Coefficients of variation for molar responses of the four compounds were <5% and for retention times <3%.

5 Application of HPLC to the Solution of Outstanding Problems in Phytoalexin Research

After establishing the chemical structure of a phytoalexin, the most important task is to assess its role unequivocally in the defence of plants against disease. Only when the phytoalexin response is proved to be a determinant of resistance of economically important crop plants will it stand a chance of attracting the attention of plant breeders. The breeder will then require rapid methods for establishing the plant's phytoalexin potential. Aside from these practical goals there are fundamental questions to be answered about the mechanism of phytoalexin elicitation and the metabolic pathways followed in their biosynthesis and degradation.

5.1 Assessment of the Role of Phytoalexins in Resistance to Microbial Attack

In order to establish unequivocally a role for a phytoalexin in the defence of a plant against a parasite it is suggested that the following criteria should be applied:

1. The phytoalexin must be synthesized and accumulate at the site of infection.
2. The phytoalexin must inhibit the growth of the parasite in vitro.

3. Inhibitory concentrations of the phytoalexin must accumulate at the site of infection at the time the parasite ceases to grow.
4. Varying the rate of phytoalexin accumulation must cause a corresponding variation in the degree of resistance.
5. Varying the tolerance of the parasite to the phytoalexin must cause a corresponding variation in the degree of virulence.

HPLC may be used to advantage in seeking to fulfil criteria (1), (3), and (4) and, when the mechanism of tolerance is the ability to degrade the phytoalexin, (5). Painstaking analysis of phytoalexin accumulation in the vicinity of the parasite must be done in order to determine whether the phytoalexin really does reach inhibitory concentrations at the time the parasite ceases to grow. Variation of the rate of phytoalexin accumulation may be achieved by using a range of cultivars differing in resistance, by the use of biosynthetic inhibitors of phytoalexins [e.g. L-2-aminooxy-3-phenylpropionic acid (Moesta and Grisebach 1982) or by employing stress conditions, e.g. drought or low oxygen tension (Wotton and Strange 1985)]. Reduction of the rate of phytoalexin accumulation should be matched by a corresponding increase in the extent of colonization of the plant by the parasite. Finally, analytical HPLC would seem to offer an ideal method of testing the ability of parasites to degrade phytoalexins, since their disappearance from incubation mixtures could be followed so easily.

5.2 Measurement of the Phytoalexin Potential of the Plant

Once phytoalexins have been proved to play an important role in the resistance of a crop plant to disease, the selection and breeding of cultivars with a well-developed phytoalexin response will become a priority. Methods of elicitation must be found and optimized which give a true measure of the phytoalexin potential of the plant and HPLC analysis of the compounds developed so that they may be done in minimal time with maximum accuracy and reproducibility. The experimenter may then embark on a survey of breeding material, testing its performance both under ideal and stress conditions. Further, he can extend his survey to wild relatives and, in doing so, may discover novel phytoalexins with effective antimicrobial properties. These data may then be used by the breeder to produce cultivars with a diverse and well developed phytoalexin response which is effective even under adverse conditions.

5.3 The Elicitation, Biosynthesis and Degradation of Phytoalexins

Important questions remain to be answered about the elicitation of phytoalexins. A bewildering array of elicitors is now known but the mechanism by which they cause the gene derepression necessary for phytoalexin synthesis is not established (Dixon 1986). A rapid HPLC assay of the phytoalexins themselves could be one contribution to such studies.

Biosynthetic studies usually involve feeding the plant with putative precursors which have been labelled (e.g. by C^{14} or C^{13}) and testing for their incorporation in the phytoalexin. It is important to separate the phytoalexin from any labelled contaminants and here HPLC would be of undoubted value. One possibility which does not appear to have been explored in such studies is the use of a radioactive detector for monitoring HPLC column effluents. Clearly, valuable clues to biosynthetic sequences could be obtained from pulse and pulse chase labelling experiments in conjunction with an HPLC equipped with such a monitor.

HPLC could also be used to provide valuable information about phytoalexin degradation. If the compound has been radiolabelled, the accumulation of degradation products as well as the disappearance of the phytoalexin could be ascertained.

6 Concluding Remarks

HPLC is an extremely versatile technique which offers the plant scientist interested in phytoalexins many opportunities. He can isolate them for chemical and toxicological studies, assay them in low concentration in plant extracts in order to establish their role in resistance and determine their biosynthetic and degradative pathways. These approaches require attention if we are to safeguard and improve the health of our crop plants.

Acknowledgments. I should like to thank the many people who wrote to me about their experience with the HPLC of phytoalexins and related compounds. In particular I am much indebted to Drs. J Goossens, G Lyon, and J Mansfield, who critically read the manuscript and made helpful suggestions.

References

Aguamah GE, Langcake P, Leworthy DP, Page JA, Pryce RJ, Strange RN (1981) Two novel stilbene phytoalexins from *Arachis hypogaea*. Phytochemistry (Oxf) 20:1381–1383

Bailey JA (1982) Mechanisms of phytoalexin accumulation. In: Bailey JA, Mansfield JW (eds) Phytoalexins. Blackie, Glasgow, pp 289–318

Bailey JA, Berthier M (1981) Phytoalexin accumulation in chloroform-treated cotyledons of *Phaseolus vulgaris*. Phytochemistry (Oxf) 20:187–188

Bhattacharyya MA, Ward EWB (1985) Differential sensitivity of *Phytophthora megasperma* f.sp. *glycinea* isolates to glyceollin isomers. Physiol Plant Pathol 27:299–310

Clouse SD, Martenson AN, Gilchrist DG (1985) Rapid purification of host-specific pathotoxins from *Alternaria alternata* f.sp. *lycopersici* by solid-phase adsorption on octadecyl silane. J Chromatogr 350:255–263

Cooksey CJ, Dahiya JS, Garratt PJ, Strange RN (1982) Two novel stilbene-2-carboxylic acid phytoalexins from *Cajanus cajan*. Phytochemistry (Oxf) 21:2935–2938

Cooksey CJ, Dahiya JS, Garratt PJ, Strange RN (1983) Sucrose: a constitutive elicitor of phytoalexin synthesis. Science 220:1398–1400

Coxon RD, O'Neill TM, Mansfield JW, Porter AEA (1980) Identification of three hydroxyflavan phytoalexins from daffodil bulbs. Phytochemistry (Oxf) 19:889–891

Dahiya JS, Strange RN, Bilyard KG, Cooksey CJ, Garratt PJ (1984) Two isoprenylated isoflavone phytoalexins from *Cajanus cajan*. Phytochemistry (Oxf) 23:871–873
Dixon RA (1986) The phytoalexin response: elicitation, signalling and control of host gene expression. Biol Rev 61:239–291
Epperlein MM, Noronha-Dutra AA, Strange RN (1986) Involvement of the hydroxyl radical in the abiotic elicitation of phytoalexins in legumes. Physiol Plant Pathol 28:67–77
Goossens JF, Van Laere AJ (1983) High performance liquid chromatography of isoflavonoid phytoalexins in French bean (*Phaseolus vulgaris*). J Chromatogr 267:439–442
Homans AL, Fuchs A (1970) Direct bioautography on thin layer chromatograms as a method for detecting fungitoxic substances. J Chromatogr 51:327–329
Keen NT (1978) Phytoalexins: efficient extraction from leaves by a facilitated diffusion technique. Phytopathology 68:1237–1239
Lee SM, Garas NA, Waiss AC (1986) High performance liquid chromatographic determination of sesquiterpenoid stress metabolites in *Verticillium dahliae* infected cotton stele. J Agric Food Chem 34:490–493
Lyne RL, Mulheirn LJ, Leworthy DP (1976) New pterocarpinoid phytoalexins of soybean. J Chem Soc Chem Commun 497–498
Lyon GD (1984) Comparison between phytoalexin concentrations and the extent of rotting of potato tubers inoculated with *Erwinia carotovora* sub sp. *atroseptica, E. carotovora* sub sp. *carotovora* or *E. chrysanthemi*. Phytopathol Z 111:236–243
Macko V, Wolpert TJ, Acklin W, Tawn B, Seibl J, Meili J, Arigoni D (1985) Characterization of victorin C, the major host-selective toxin from *Cochliobolus victoriae:* structure of degradation products. Experientia 41:1366–1370
Mansfield JW (1982) The role of phytoalexins in disease resistance. In: Bailey JA, Mansfield JW (eds) Phytoalexins. Blackie, Glasgow, pp 253–288
Mansfield JW, Porter AEA, Smallman RV (1980) Dihydrowyerone derivatives as components of the furnoacetylenic phytoalexin response of tissues of *Vicia faba*. Phytochemistry (Oxf) 19:1057–1061
Mansfield JE, Barlow YM, Porter AEA (1984) Phytoalexin production by *Vicia faba* – a model system for the study of gene expression. In: Chapman GD, Tarawali SA (eds) Systems of cytogenetic analysis in *Vicia faba* L. Nijhoff, Junk
Moesta P, Grisebach H (1982) L-2-Aminooxy-3-phenylpropionic acid inhibits phytoalexin accumulation in soybean with concomitant loss of resistance against *Phytophthora megasperma* f.sp. *glycinea*. Physiol Plant Pathol 21:65–70
O'Neill TM, Mansfield JW (1982) Mechanisms of resistance to *Botrytis* in narcissus bulbs. Physiol Plant Pathol 20:243–256
Osswald WF (1985) High performance liquid chromatography of isoflavonoid phytoalexins from soybean (*Glycine max* L. Harosoy 63). J Chromatogr 333:225–230
Paxton J (1981) Phytoalexins – a working redefinition. Phytopathol Z 101:106–109
Porter AEA, Smallman RV, Mansfield JW (1979) Analysis of furanoacetylenic phytoalexins from the broad bean plant by high performance liquid chromatography. J Chromatogr 172:498–504
Rahe JE, Arnold RM (1975) Injury-related phaseollin accumulation in *Phaseolus vulgaris* and its implications with regard to specificity of host-parasite interaction. Can J Bot 51:921–928
Ride JP, Drysdale RB (1972) A rapid method for the chemical estimation of filamentous fungi in plants. Physiol Plant Pathol 2:7–15
Rohringer R, Kim WK, Samborski DJ, Howes NK (1977) Calcofluor: an optical brightener for fluorescence microscopy of fungal plant parasites in leaves. Phytopathology 67:808–810
Seitz LM, Sauer DB, Burroughs R, Mohr ME, Hubbard JD (1979) Ergosterol as a measure of fungal growth. Phytopathology 69:1201–1203
Sims JJ, Keen NT, Honwad VK (1972) Hydroxyphaseollin, an induced antifungal compound from soybeans. Phytochemistry (Oxf) 11:827–828
Still WC, Kahn M, Mitra A (1978) Rapid chromatographic technique for preparative separations with moderate resolution. J Org Chem 43:2923–2925

Strange RN, Ingham JL, Cole DL, Cavill ME, Edwards CE, Cooksey CJ, Garratt PJ (1985) Isolation of the phytoalexin medicarpin from leaflets of *Arachis hypogaea* and related species of the tribe Aeschynomeneae. Z Naturforsch 40:313–316

Wotton HR, Strange RN (1985) Circumstantial evidence for phytoalexin involvement in the resistance of peanuts to *Aspergillus flavus*. J Gen Microbiol 131:487–494

Yoshikawa M, Yamauchi K, Masago H (1978) Glyceollin: its role in restricting fungal growth in resistant soybean hypocotyls infected with *Phytophthora megasperma* var. *sojae*. Physiol Plant Pathol 12:73–82

Analysis of Lipids by High Performance Liquid Chromatography

L. A. SMITH and G. A. THOMPSON, JR.

1 Introduction

High performance liquid chromatography (HPLC) of natural lipids has been remarkably slow to develop as an analytical tool. The absence of easily detectable functional groups on most lipids makes them difficult to measure quantitatively using the more common types of detectors. As a result, many of the published methods for underivatized lipid HPLC are semiquantitative at best and are primarily designed to separate rather than quantify the individual components. The structures of some lipids are amenable to derivatization, yielding light-absorbing products that may be quantified on a molar basis. Applications of this type have been quite successful, although extreme care must be taken to avoid degradation or selective losses of some lipid species during the derivatization steps.

Lipid HPLC applications may be roughly divided into two categories: the separation of lipid classes from each other and the separation of individual molecular species within a lipid class. Generally speaking, the lipid class separations are usually achieved by absorption chromatography on silica gel columns, while molecular species are resolved most successfully on reverse phase columns.

The newer methods for resolving lipid classes provide much cleaner separations than could be obtained using more classical low pressure silica gel column chromatography, and can be used with much larger amounts of lipid than are conveniently separable by thin layer chromatography. So apart from the considerably higher capital investment on instrumentation, HPLC is a distinct improvement for the isolation of lipid classes. However, the most spectacular achievements in lipid HPLC have come in molecular species separations. Here the most apt comparison is with gas chromatography, but unlike gas chromatography, which can only resolve volatile mixtures, reverse phase HPLC has the potential of separating any group of closely related lipids. It provides us, then, with a new dimension in lipid analysis, making possible studies that could not have been attempted previously.

Perhaps the best illustration of this new capacity is the ease with which membrane polar lipid classes can be resolved into their individual molecular species. In the case of phospholipids, this analysis has been possible for some time using gas chromatography, but it may only be accomplished by first hydrolyzing the purified phospholipid with phospholipase C to remove the polar head group and then converting the resulting diacylglycerol mixture to trimethylsilyl or other volatile derivatives prior to chromatographic analysis (Lynch and Thompson 1982). Besides being inconvenient and time-consuming, there is a real likelihood of incurring selective losses during this preparation. In contrast, HPLC can separate intact underivatized phospholipid molecular species (Smith et al. 1985). Furthermore, it cleanly resolves the molecular species of other types of polar lipids, such as mono- and digalactosydiglycerides, which cannot with present techniques be transformed into volatile compounds.

2 Instrumentation

2.1 Pumps, Injectors, and Columns

HPLC analyses of lipids can be accomplished with the basic HPLC component system consisting of a pump, injector, column, detector, and recorder. Special consideration should be given to some of these components when they are to be used for lipid analysis, while others may be selected following general guidelines (e.g., Booth and Quinn 1983). For example, any commercial HPLC pump or injector can be utilized for lipid analyses as long as their capabilities for the desired type of chromatography (i.e., microbore, preparative, gradient, etc.) are taken into account. Several characteristics of columns, however, should be tailored especially for the intended lipid analysis. The nature of the separation dictates such parameters as the stationary phase type (e.g., normal or reverse phase), column diameter, and particle size. These latter two characteristics are particularly important, as they will determine the maximum limits of sample loading and flow rate. If part of the goal, for example, is collecting lipids as well as separating and detecting them during the HPLC run, then a column that can handle the necessary sample load must be selected. Furthermore, if the detector utilized destroys the lipid during the detection process, collection of samples can only be accomplished by diverting a portion of the column eluate. Eluate stream splitting requires the use of larger sample loads. Large sample loading could also be necessary where enough material must be collected to obtain an accurate measurement of radioactivity in low specific radioactivity lipids. As long as these aspects are taken into account, then columns obtained from any of the various commercial manufacturers should be suitable.

2.2 Detectors

One of the most critical components of HPLC systems for lipid analyses is the detector. Over the last decade, monitoring peak elution via the absorption of ultraviolet light has been by far the method of choice. The popularity of this detection mode is primarily due to its sensitivity towards a large number of lipids in the 200–210 nm range. Detection in this way is nondestructive, facilitating the collection of separated samples. However, there are a number of serious limitations to UV detection in lipid HPLC analyses. Detection of natural lipids without derivatization relies on the presence of UV-absorbing functional groups, most commonly, unsaturated acyl chains. As the degree of unsaturation (and therefore

Table 1. Characteristics of microparticulate packings for aqueous SEC of polypeptides

	Company	Packing base	Particle size (micron)	Pore size (Å)	pH range	Recommended M_r range	Optimal M_r range	
							0.1% SDS	6 M Gu · HCl
TSK G 2000SW[a]	Toyo Soda	Silica	10	130	2.5–8	5×10^3–3×10^4	N.R.[c]	10^3–10^4
3000SW[a]			10	240	2.5–8	3×10^4–5×10^5	10^4–6×10^4	10^4–7×10^4
4000SW[a]			13	450	2.5–7.5	above 5×10^5	above 6×10^4	above 7×10^4
TSK G 3000PW	Toya Soda	Hydroxylated polyether	13	300	2–12	5×10^2–1.5×10^5 [b]	–	
4000PW			13	400		–		
5000PW			17	500		–		
Synchropak APC 100	SynChrom	Silica	5	100	2–8	3×10^3–5×10^4	[d]	[d]
200			5	200		–[d]		
300			5	300		–		
500			10	500		–		
Protein column I 60	Waters	Silica	10	60	2–8	–[d]	[d]	[d]
I 125			10	125		–		
I 250			10	250		–		
Zorbax GF 250	Dupont	Zirconium-treated silica	4	–	2–8.5	4×10^3–4×10^5		
Fractogel TSK HW 40	Merck	Polyvinylacetate	52–63 (F)	–	1–14	10^2–10^4		
HW 45			25–40 (S)	–		10^3–7×10^5		
Superose 6	Pharmacia	Agarose	13 ± 2	–	1–14	5×10^3–5×10^6		
12			–		10^3–3×10^5			

[a] Molecular weight range indicated represents the optimal range; the actual separation range is larger.
[b] Swergold and Rubin (1983).
[c] N.R. not recommended.
[d] Reliable data not available but an estimate may be obtained by comparison of the pore sizes with the TSK-SW series.
Data obtained from manufacturer's information and literature except where indicated.
The Synchropak GPC series is also available from Brownlee Labs as Aquapore OH; TSK-SW series can be purchased from BioRad Labs, Varian Assoc., Beckman-Altex, Inc., and LKB Instruments, respectively, as the Bio-SilTSK, Micropak TSK, Microspherogel TSK and Ultropak TSK series.

under nondenaturing conditions and the TSK-SW type packings in the presence of 0.1% SDS or 6 M guanidine · HCl. Indeed, since size exclusion depends on hydrodynamic volumes of the polypeptides, which increase considerably upon addition of denaturing agents to the solution, the separation range of a specific type of column will shift to a lower molecular weight region in the presence of these denaturing additives as compared to the regular buffer. As a general rule, provided that nondenaturing conditions are used, polypeptides of 50 kd and smaller are most successfully separated on packings with 100 Å pore size particles. By increasing the SDS concentration to 0.2 or 0.3%, the effective molecular weight separation range of polypeptides on a TSK 3000 SW column can even be extended down to 10^3 dalton. A TSK 2000 SW column operated with 0.3% SDS in the mobile phase was used to separate medium-sized peptides ranging in molecular weight from 500 to 6000 (Mabuchi and Nakahashi 1981). The separation range of a system may also be increased by connecting two columns with different pore size in series; the one with the largest pore size is placed first (see also Sect. 2.2.2).

The proper choice between a TSK-SW- or PW-type packing depends on the purpose of the experiment and the molecular weight distribution of the polypeptides. The advantages of PW packings are a wider molecular weight separation range from a single column, which is useful in molecular weight determinations, and possible applications for small peptides (100–1000 dalton) and very large macromolecules (proteins and DNA fragments larger than 200 kd). SW type columns are generally better for proteins in the 10 to 200 kd molecular weight range because they offer higher efficiencies and resolution (Alfredson et al. 1982). Particular applications of both types of columns will be discussed later in the text.

2.2 Operational Parameters and Resolution

Although the performance of SE-HPLC is considerably superior to its open column counterpart, the resolution is still rather limited. Under optimal conditions, two polypeptides with molecular weights that differ by more than 1.5-fold (e.g. 30 kd and 45 kd) can be separated (Pfannkoch et al. 1980). We will examine three important operational variables and how they affect resolution.

2.2.1 Mobile Phase

As pointed in Sect. 2.1.1, salt concentrations and certain organic modifiers are important to eliminate nonspecific surface interactions and assure optimal SE separations. Additives of different kinds to the mobile phase are also necessary when the researcher wants to prevent aggregation of proteins or to preserve biological activity of enzymes. A summary of most common mobile phases and additives is given in Table 2. Their use will be discussed in more detail in the practical applications section (2.3). A possible negative effect of additives on resolution may occur by way of increased viscosity of the mobile phase. This influences the normal diffusion of the solutes in and out of the particles, resulting in bandspreading and decreased resolution. Viscosity effects will occur with such additives as urea,

	Eliminate aspecific interaction		M_r determination	Solubilize disaggregate	Biolog. activity	Structural studies	Selected references
	Ionic	Hydro-phobic					
Buffers/ionic strength							
Sodium phosphate buffer	$\geqq 0.1$ M +	$\geqq 0.3$ M −				+	1, 2
Sodium acetate buffer	$\geqq 0.1$ M +	$\geqq 0.3$ M −				+	2
(NaCl, KCl)	$\geqq 0.1$ M +	$\geqq 0.3$ M −				+	3, 4
Organic solvents							
Ethanol		+	+	(+)	> 10% −		4
Methanol		+			−		5
Acetonitrile in							
0.1% TFA[a,b]	+	+	+	(+)	−	+	6
0.1% H_3PO_4[a]	+	+	+	(+)	−	+	6
0.1% HCl[a]	+	+		+	−	+	7
TEAP pH < 3[c]	+	+		(+)	−	+	8
Organic additives							
Guanidine-HCl (6 M)		+	+	+	−	+	9, 10
Urea (6 M)		+		+	−	+	11
SDS (0.1–0.3%)	−	+	+	+	−	+	1, 12, 13
Deoxycholate (0.1–0.25%)		+		+	+		14, 15
Triton X-100 (0.05–0.1%)				+	+		14
Octylglucoside		+		+	+		16
Glycerol					+		17
Polyethylene glycol		+			+		18

The signs + and − indicate special advantages or disadvantages, respectively, of certain ions, solvents or additives for a particular application. Blanks indicate no obvious positive or negative effect or no data found in literature.

References: 1, Takagi et al. (1981); 2. Jenik and Porter (1981); 3. Kato et al. (1980a); 4. Hefti (1982); 5. Barth (1982); 6. Swergold and Rubin (1983); 7. Welling et al. (1985); 8. Rivier (1980); 9. Ui (1979); 10. Kato et al. (1980c); 11. Wehr et al. (1982); 12. Imamura et al. (1979); 13. Kato et al. (1980b); 14. Ludi and Hasselbach (1984); 15. Lambotte et al. (1984); 16. Delucas and Muccio (1984); 17. Gruber et al. (1979); 18. Fuhrer and Evans (1982).

[a] Used in combination with a TSK-PW type column.

[b] TFA: Trifluoroacetic acid.

[c] TEAP: Triethylamine phosphate buffer.

guanidine hydrochloride, and glycerol, and are even more pronounced at lower temperature. In these cases, lower flow rates are particularly recommended. Finally, changes of pH at low ionic strength may influence the resolution between a discrete set of polypeptides (Sect. 2.1.1).

2.2.2 Sample

Injection volume and sample concentration are very important in SEC. For the best resolution, injection volume on a (7.5 × 300 mm) column should not exceed 100 μl and the total polypeptide mass for this volume should be less than 1 mg (final concentration 1%); for a 4.1 × 300 mm column the appropriate volumes and masses are 30 μl and 300 μg, respectively. However, sample concentrations should be kept below 0.5% for proteins larger than 80 kd. Solutions of high molecular weight proteins are more viscous than solutions of smaller molecules and increased sample viscosity results, as in case of the mobile phase, in decreased resolution. For good separations, care should be taken that sample viscosity is roughly equal to at most 50% more than the mobile phase viscosity. Also, if two columns of different pore size are connected in series, they should be in order of decreasing pore size to separate and dilute out the high molecular weight polypeptides first and avoid overloading and bandspreading. Load and injection volume for preparative SE-HPLC will be discussed further in Sect. 2.4 of this review.

2.2.3 Mobile Phase Velocity

Correct mobile phase velocity is critical for high resolution separation of polypeptides. Flow rates for a 7.5 × 300 mm column should be below 1 ml min^{-1} (mobile phase velocity of 2 cm min^{-1}) to give fair efficiencies. However, if the column is operated in a common buffer system of low viscosity, better resolutions are obtained at lower flow rates such 0.5 ml min^{-1} or less. This correlation is more pronounced for bigger proteins and with increased mobile phase viscosity; in both cases the flow rate should be lowered correspondingly. For the separation of proteins larger than 100 kd on a 7.5 × 300 mm TSK 4000 SW column in the presence of 0.1% SDS, we obtained optimal resolutions at a flow rate of 0.3 ml min^{-1}.

2.3 Practical Applications

2.3.1 Molecular Weight Determination

Since SE-HPLC separates polypeptides based on hydrodynamic volumes, it offers a possible alternative for SDS-gelelectrophoresis to determine molecular weights. However, two problems, one inherent to the nature of the protein the other to the column support, cause a major obstacle for reliable results. First, SE separations depend on the shape of the molecules; therefore, shape should be as uniform as possible and linearly related to the molecular weight. This uniformity can be most easily obtained by complete denaturation and unfolding with, for example, SDS or guanidine · HCl; in addition, some researchers like to reduce and alkylate the proteins first. The second problem is nonspecific surface interactions;

possible ways to eliminate those are discussed in Sect. 2.1.1 and summarized in Table 2. Under optimal conditions, a linear relationship between log M_r of a protein and its retention time on a SE column is obtained. Commonly used mobile phases for polypeptides are 6 M guanidine · HCl in 10 mM sodium phosphate buffer (Ui 1979; Richter et al. 1983) and 0.1% SDS (Imamura et al. 1979) to 0.3% SDS (Mabuchi and Nakahashi 1981). SDS-protein complexes carry a negative charge and are therefore ionically excluded from the silica matrix; for that reason SDS is always used in combination with 0.1 to 0.15 M sodium phosphate buffer. Linear calibration curves for standard polypeptides were also obtained on a Waters I-125 protein column with 20% ethanol/0.32 M NaCl (Hefti 1981) or 15 to 30% acetonitrile in triethylammonium phosphate buffer, pH 2.25 (Rivier 1980) as mobile phases and on a TSK G 3000 PW column using 36 to 45% acetonitrile in 0.1% TFA or 0.1% H_3PO_4 (Swergold and Rubin 1983). Organic solvent concentration proved to be very critical in these experiments, as was acidic pH with the TSK-PW column.

2.3.2 Preservation of Biological Activity

In order to maintain the binding of tritiated estradiol to its receptor during analysis, chromatography was carried out on a Synchropak GPC-100 (4.6 × 250 mm) column with 10% glycerol in phosphate buffer (pH 7.2) as mobile phase, at 0 °C and with a flow rate of 80 μl m^{-1} (Gruber et al. 1979).

Lambotte et al. (1984) used 0.25% deoxycholate in 150 mM NaCl/20 mM sodium phosphate buffer (pH 7.4) with a Waters I-125 column to fractionate membrane glycoproteins from mastocytoma cells. Fractions were analyzed and biological activity observed by an in vitro assay involving T-lymphocyte cell culture.

Lymphotoxins, a multicomponent immunological hormone family with molecular weights of 20 to 56 kd, were partially purified by Fuhrer and Evans (1982) with nearly complete recoveries of biological activity, by passing through two Waters I-125 (7.8 × 300 mm) columns equilibrated and eluted with 0.1% polyethylene glycol in 0.1 M sodium phosphate buffer pH 7.4.

Partially affinity purified murine platelet-derived growth factor (PDGF) receptor preparations have been purified to homogeneity using two TSK 4000 SW (7.5 × 300 mm) columns and 0.05% Triton X-100 in 0.1 M sodiumphosphate pH 7.0, with retention of PDGF binding activity and PDGF-dependent tyrosine kinase activity (D. Woo and P. Tempst, unpublished observations 1985).

2.3.3 Selected Applications

The use of guanidine · HCl in the SEC of the human platelet-derived growth factor, a "sticky" protein that adheres easily to surfaces and was isolated from human platelets, the epidermal growth factor receptor, a membrane glycoprotein (M_r 160 kd) and CNBr fragments of the latter, was successfully demonstrated by Waterfield and coworkers (Waterfield et al. 1983; Downward et al. 1984). For this purpose, TSK 3000 SW (for PDGF and EGF-R CNBr) and TSK 4000 SW columns (7.5 × 600 mm) were used with 6 M guanidine · HCl in 0.1 M potassium phosphate (pH 4.5). These purifications allowed subsequent protein sequence analysis.

SE-HPLC on a TSK 4000 SW column in the presence of 0.1% SDS was used to fractionate Sendai virus envelope glycoproteins (Welling et al. 1984). The buffer was 50 mM sodium phosphate (pH 6.5) and the Triton X-100-extracted proteins were boiled for 2 min in 4% SDS before injection. A similar approach was taken in our laboratory for the final purification of mouse PDGF-receptor. It was accomplished by two subsequent passages over the same type of column with 0.1% SDS in 0.1 M sodium phosphate (pH 7.0) and the protein mixtures were incubated with 2% SDS at 37 °C for 10 min prior to injection. The separation profiles and analytical SDS-gelelectrophoresis of collected fractions are presented in Fig. 7.1. SDS has the advantage that, unlike guanidine · HCl, it does not have to be removed prior to SDS-gelelectrophoresis.

Octylglucoside as an additive to the SEC mobile phase proved to be very useful in the purification of bovine rhodopsin, a hydrophobic glycoprotein with a linked chromophore and a M_r of 41 kd (Delucas and Muccio 1984). After solubilization of the retinal membrane proteins in a Tris chloride buffer containing 30 mM n-octylglucoside, the mixture was chromatographed on a TSK 3000 SW column with 30 mM octylglucoside and 150 mM NaCl in 60 mM phosphate buffer (pH 6.5) as a mobile phase.

In order to determine and obtain reproducible molecular weight distribution patterns of hydrolyzed plant proteins (soy, potato, cotton seed, corn gluten, wheat bran and germ), Barth (1982) investigated mobile phase composition for SE-HPLC with respect to performance and recovery. Synchropak silica-based columns with mobile phases containing either 1% SDS or 30% methanol in addition to 50 mM Na_2SO_4 gave the most satisfactory results. This technique finds its application in the quality control of hydrolyzed protein in the food industry.

TSK-PW type columns have been used in the isolation of some hydrophobic glycoproteins and very large polypeptides. Welling and coworkers (1985) separated Sendai virus envelope proteins on a TSK 4000 PW column with 45% acetonitrile in 0.1% HCl as mobile phase. This allowed UV detection at 210 nm and the collected fractions to be chromatographed easily on a reversed phase column. Chromatography of chlorocruorine, a huge protein of $M_r = 2.9 \times 10^6$, and other large proteins was successfully accomplished as a TSK 5000 PW column equilibrated and eluted with 0.1 M KCl in 10 mM phosphate pH 7.0 (Himmel and Squire 1981).

2.4 Large-Scale Separations

Regular size analytical SE-HPLC Column (4.1 to 7.5 × 300 mm) are also used by most researchers for preparative purposes. However, if the sample volume and protein mass exceed the values given in Sect. 2.2.2, columns should be scaled up or multiple injections considered. The latter possibility is sometimes easy and is definitely the cheapest option. The simplest way to scale up is to put two columns in series. The sample volume can then be increased by as much as 50%; however, peaks will be broader. The series can be extended, but this is a very costly procedure compared to making or buying a preparative column. These usually measure 2×100 cm or larger, and loads of 100 mg in 10 to 20 ml can be applied. For large loads, bigger particles (20–40 micron) give as good results as small particle packings and are considerably cheaper. Medium performance gels such as Superose (Pharmacia) and Fractogel TSK (Merck) also offer cheaper alternatives and can be used in glass columns. For larger columns, the flow should be scaled up in proportion to the increase in cross-sectional area. A column with an inner diameter of 2 cm can be used with flow rates up to 10 ml min^{-1}.

2.5 Maintenance

If a SE system is to be stored for a short period of time, it is a good idea to pump several column volumes of distilled water through the system. It also protects all metal parts of the system from corrosion in cases where salts and especially halides are used. In a longer period of storage, the columns should be kept in 0.1% sodium azide or 10 to 100% methanol. This prevents microorganisms from growing and methanol has the additional advantage that it can desorb adsorbed polypeptide material. However, if the column is to be used for smaller polypeptides (1–10 kd), storage in NaN_3 is recommended. According to Link et al. (1984), a SE column needs to be presaturated with some standard peptide to avoid peptide adsorbtion and ensure good recovery of peptides subsequently chromatographed; methanol storage renews the binding capacity of this type of column for peptides.

Because of adsorbtion of polypeptides and particulate or any other material at the top of the column, it is highly recommended to use a guard column, especially in cases of preparative separations. If the performance of a system decreases and the backpressure goes up, one can clean the guard column by pumping mobile phase through it in the reverse orientation and/or clean the top frit of the column by sonicating in 3 M HNO_3 or in 70% acetic acid saturated with guanidine · HCl. In case of adsorbed polypeptides, a few gradients of propanol (0–60%) in 0.1% TFA may be run over the guard column.

SDS is so hydrophobic that, when used as additive to a mobile phase, it sticks very strongly to the column supports and is almost impossible to remove. It is therefore recommended not to use the column afterwards with any different mobile phase and to reserve it exclusively for use with SDS. If a system has to be changed from use with SDS to guanidine · HCl or vice versa, switch columns but also rinse the complete system carefully with water, since mixing SDS and guanidine · HCl causes precipitates that will plug injection valves.

3 Ion Exchange HPLC

As mentioned before, the charge of a polypeptide in solution depends on the pH of the solution and its specific titration curve. Net charge is positive, or negative at a pH below, or above the isoelectric point (pI). Ion exchange chromatography

(IEX-C), which also has long been used by biochemists in open column systems, is based on the electrostatic surface interactions between polypeptide solutes and charged ligands on a solid support. The ligand charges can be positive or negative, which enables ionic binding of anions, or cations; the packings are therefore called anion exchangers (AEX) and cation exchangers (CEX) respectively. Adsorbed polypeptide solutes can be desorbed by either changing the pH (raise in case of CEX, decrease for AEX), or increasing the ionic strength of the mobile phase. Besides general superiority in speed and resolution, IEX-HPLC offers an additional advantage over classical low pressure, open column systems since silica-based packing do not shrink or swell upon ionic strength changes as is the case for cellulose or agarose supports.

Different brand columns in combination with such variables as mobile phase pH and counterions, flow rate, and gradient slope allow separation of closely similar polypeptide from highly complex mixtures. IEX-HPLC is widely used for the separation of proteins, especially since the recoveries are generally high and the biological activity is usually maintained under standard operating conditions. A selected number of applications, taken from literature, will be discussed with respect to the nature of the protein (hydrophobicity, size) and the characteristics of the mobile phase.

Since gradient elution of IEX-columns is general, an IEX-HPLC system must contain two pumps, a gradient mixer and a controller unit. Eluent detection is usually performed at 280 mm. Because the mobile phase may contain high concentrations of salts, special care is necessary not to leave the mobile phase in the HPLC system. Corrosion of the stainless steel parts may occur, especially in the case when halides are used, and crystallization of salts is also possible. Therefore, it is recommended to replace the buffer with distilled water every night. Columns should also be stored in distilled water.

3.1 Columns

High performance IEX packings usually consist of silica based porous spherical particles (5–10 micron) with a polymer surface coating that holds the charged ligands. This provides a pressure resistant packing, stable in the pH range 2–8. An organic based IEX packing is available from Pharmacia under the trade name Mono beads; these packings can be used with mobile phase pH up to 12. Most packings exhibit slight hydrophobic surface interaction with a limited number of polypeptide solutes; this nonideal surface interaction can usually be eliminated by addition of up to 30% organic solvent to the mobile phase. Several general anion- and cation exchanger ligands exist, and in each case one can distinguish between weak and strong ion exchangers. As for classical IEX media, strong ion exchangers have constant capacities over the complete operating pH range, while for weak ion exchangers, especially of the DEAE type, pH changes result in an altered number of charged groups and thus in capacity. The choice between a cation- and anion exchanger depends on the pI of the polypeptide. As a general rule, CEX-HPLC can be used for proteins with pI's > 5 and AEX-HPLC for proteins that have pI's < 7.

Since IEX chromatography is a surface interaction process, and because 95% of the surface area is inside the particles, solutes should be able to diffuse freely through the pores. If the pore size is too small, large polypeptides cannot get inside the particles and this will decrease loading capacity and resolution dramatically. A pore size of 100 Å is efficient for polypeptides with M_r up to 30–50 kd, while larger proteins with M_r up to 200 kd should be chromatographed on supports with 300 Å pore size particles. Thus selection of a proper high performance IEX column depends largely on the pI and M_r of the protein(s) to be isolated. Several HPLC-packing manufacturers offer prepacked IEX columns. A survey of recent literature indicates that Mono Q and Mono S (Pharmacia), TSK (Toya Soda), and Synchropak (SynChrom) packings are most popular among protein chemists today.

Since IEX-HPLC is primarily based on a displacement chromatography mechanism, column length is more important for total capacity and sample loads, but not for resolution. Vanecek and Regnier (1980) determined that columns of 5 cm length had 10% protein load capacity of a 25 cm column, but the resolution in case of small loads was still 75% compared to the longer column. Thus for microscale protein separations, column length should be kept to a minimum since it will not affect resolution and polypeptides will be eluted in smaller volumes resulting in increased detection sensitivity. Brownlee Labs offers guard columns (4.6 × 30 mm) with 300 Å pore size high performance IEX packings that are especially suited for this purpose.

3.2 Operational Parameters and Resolution

Resolutions of two polypeptide solutes in gradient elution is usually expressed as:

$$R_S = \frac{2(Ve_2 - Ve_1)}{\Delta Ve_2 + \Delta Ve_1} \quad \text{(Regnier 1984)},$$

where Ve_1 and Ve_2 are the elution volumes of solutes 1 and 2, respectively, and ΔVE_2 and ΔVe_1 are the corresponding peak widths in volume. Thus any change in parameters that will result either in greater separation or sharper peaks will increase resolution. Depending on the nature of the proteins, alteration of variables such as pH, counterions, ionic strength, flow and gradient slope will influence retention and peakwidths in a very specific way, either synergistically or antagonistically with respect to resolution. In practice, a suitable set of conditions has to be searched for empirically.

3.2.1 Mobile Phase

Choice of mobile phase pH depends on the use of an anion- or cation exchanger and also on the pI's of the proteins to be separated. If the mobile phase pH equals the pI of a protein, the net charge will be zero, and the protein while still retained by the ion exchange support should be easily eluted at low ionic strength. However, discrepancies have frequently been observed, as has also been the case with open column IEX chromatography, in cases where the protein is still adsorbed rather strongly at pH values near the pI. This can be explained by assuming that the charges are distributed assymetrically on the protein molecule, so that even if the net charge = 0, there may exist regions of the protein that still carry a substantial positive or negative charge (Kopaciewicz et al. 1983). Because of the tremendous differences in pI's, charge distributions and surface accessibility of charged domains of polypeptides, manipulation of mobile phase pH offers the greatest selectivity in IEX-HPLC of these molecules. Buffer systems popular for use with anion exchangers are Tris or alkylamine (e.g., Triethylamine) buffers; acetate and phosphate buffers are often used with cation exchangers.

Adsorbed polypeptide solutes are desorbed from the matrix by increasing the concentration of counterions in the mobile phase. This can be done by either gradually increasing ionic strength of the buffer salts or by a salt gradient in a low ionic strength buffer. In either case, starting ionic strength usually varies from 20 to 50 mM and is raised to 0.2–0.5 M final concentration. Ionic strength has to be higher to elute polypeptides from high performance IEX packings compared to open column IEX resins, because of their considerably higher ligand density. Counterions act mainly as displacers of the polypeptides because of their affinity for the matrix surface ligands but can also act on the protein solutes and partially neutralize their charges. Protein retention is greatly affected by the choice of displacing counterions because of their different affinities for the matrix ligand; protein counterions are of minor influence for retention. The relative effects of a number of counterions on protein retention in IEX-HPLC are given in Table 3. It should be noted, however, that these series represent the average for a number of proteins, but may be different for any individual protein. In addition, counterions also affect protein peakwidth in a specific way (Gooding and Schmuck 1984). Thus, counterion selection offers one way of varying selectivity for a series of separations on a single column.

No special precautions against loss of biological activity are necessary in the majority of IEX-HPLC experiments. Addition of detergents to the mobile phase that carry no net charges or no charge at all, such as Triton X-100, CHAPS and Berol to keep hydrophobic proteins solubilized, does not interfere with the mode of separation. However, increased salt concentration lowers the critical micelle concentration (CMC) for those detergents. As a result, micelle formation may be promoted at a certain point of the gradient and dramatic baseline shifts, due to refractive index differences, will then occur when UV detection is used. Investiga-

Table 3. Effects of counter ions on the retention of proteins during IEX-HPLC

	Displacers	Protein counter ions
AEX-HPLC	Citrate > sulfate > phosphate > chloride > formate > acetate	$Ca^{2+} > Mg^{2+} > NH_4^+ > Na^+ > K^+$ (as chloride salt)
CEX-HPLC	$Ba^{2+} > Ca^{2+} > Mg^{2+} > NH_4^+ > K^+ > Na^+$	Acetate > chloride > sulfate > phosphate (as sodium salt)

Counter ion series are in sequence (left to right) of increased protein retention. Data are from Kopaciewicz et al. (1983), Gooding and Schmuck (1984), and BioRad product information.

tion of the changes in CMC of a detergent with a particular salt should be done in preparation to such an IEX-HPLC experiment to avoid disturbing effects (Karsnas et al. 1983). In many cases the addition to the mobile phase buffer solutions of up to 30% polar organic solvents such as acetonitrile or methanol, facilitates elution and improves resolution presumably because they eliminate aspecific hydrophobic interactions. The use of urea to improve recoveries has also been reported. With all mobile phase additives, care should be taken that the ultimate viscosity is not too high, since this will cause increase of peak widths. Examples of the use of additives will be discussed in the practical applications section.

3.2.2 Mobile Phase Velocity

If the flow is too fast, solutes cannot diffuse sufficiently in and out of the support particles, especially in the case of big proteins. Thus an increase of flow rate will result in decrease of resolution. The minimal practical flow for a 4.1×300 mm column is 0.25 ml min^{-1}, although for most applications a flow of 0.5–1 ml min^{-1} is most convenient with respect to duration of the experiment. The size of proteins, however, has to be considered an important factor when choosing the mobile phase velocity.

3.2.3 Gradient Slope

The increase of ionic strength as a function of time is commonly referred to as the gradient slope of an IEX chromatography experiment. A shallow gradient slope will result in longer separation times but will tend to give better resolution than a steep gradient. Therefore, in the separation of proteins with nearly the same chromatographic properties, isocratic elution with a suitable eluent may be optimal. With shallow gradients or isocratic elution, peak widths will increase. As a rule of thumb, gradient slopes should be taken in such a way that 20 to 40 column volumes are required to reach the final ionic strength necessary for elution of all solutes. Thus, when IEX columns are operated at lower flow rates, gradient slopes should be decreased accordingly.

3.2.4 Sample and Load

Because of the charge-based adsorbtion/desorbtion mechanism, the capacity of an IEX-HPLC column is about 100 times higher than for a SE-HPLC column. Overloading will, of course, increase the peakwidths of eluted peptides or proteins but depending on the complexity, 0.5 to 10 mg of protein mixture can be separated on a 4.1×300 min column packed with 300 Å pore size particles. Since this exceeds the amount of material routinely available to protein chemists, IEX-HPLC can be a useful preparative tool. The volume of sample loaded is of very little importance as long as the ionic strength and viscosity are sufficiently low because the protein is adsorbed out of solution onto the top part of the column packing. Therefore, a diluted sample is recommended. For scaled-up operations, most of the considerations of SE-HPLC apply also for IEX-HPLC. However, in this case, columns with larger inner diameter (high cross-sectional area) and according higher flow rates are desirable; column length is of secondary importance.

3.3 Selected Applications

IEX-HPLC is a powerful technique for the isolation of proteins but has not found wide use in chromatography of peptides because of the superiority of reversed-phase HPLC in the separation of peptides. However, IEX-HPLC has been used as part of multi-dimensional peptide mapping and isolation where it frequently is a very useful first fractionation step; partially purified peptide mixtures are then adsorbed onto a RP-column, the salt removed and the peptides separated (see Sect. 5). Here we want to mention some recently published IEX-HPLC experiments that are interesting because they deal with either complex mixtures and very hydrophobic or very big proteins such as membrane receptors.

The Mono Q (Pharmacia) anion exchange column has found widespread application. Two different Sendai virus envelope glycoproteins, hemagglutinin-neuramindase protein (M_r 66 kd) and a fusion protein (M_r 65 kd), could be separated completely when using a 0–0.5 M NaCl gradient in 20 mM Tris · HCl pH 7.8 supplemented with 0.1% Triton X-100 to keep the proteins in solution; immunological reactivity with antibodies was retained (Welling et al. 1984). Chromatography of a 420 kd glycoprotein, thrombospondin, was carried out with success by employing a 0–1 M NaCl gradient in 20 mM triethanolamine · HCl, pH 7.4 (Clezardin et al. 1984); McGregor et al. (1985) reported separation of four major platelet membrane glycoproteins in essentially the same system but with 0.5% Berol added to the mobile phase. During final purification of NADPH-dependent cytochrome P-450 reductase (78 kd) and aromatase (52 kd) on a Mono Q column, proteins were kept in solution and the biological activity, upon recombination of the two components, was preserved by adding 0.5% CHAPS, 10% glycerol, and 0.2 mM EDTA to the Tris acetate buffered mobile phase (Muto and Tan 1985).

Ribosomal proteins of *E. coli* are very diverse in hydrophobicity and size. Mixtures of 21 or 26 proteins, most of them with M_rs <25 kd, were chromatographed on a TSK IEX-530 CM column to isolate individual protein species under native conditions. A gradient of 0–0.3 M KCl was applied in a 10 mM sodium phosphate (pH 7.0) mobile phase containing 5 M urea to guarantee sufficient recoveries of the proteins (Kamp et al. 1984).

The enormous resolving power of a properly selected IEX-HPLC system was demonstrated by Kumagaye et al. (1985) when they separated two synthetic variants of parathyroid hormone (84 amino acids), only different in one Asp→Asn substitution, on a TSK CM-2 SW cation exchanger. A linear 20–100% gradient over 30 min at 42 °C was employed. Buffer A consisted of 10% acetonitrile in 20 mM sodium phosphate buffer pH 6.0; buffer B was the same but contained 0.5 M NaCl and the pH was 7.0.

One of the subunit proteins of the chloroplast coupling complex is so hydrophobic that it is only extractable and soluble in pure or high concentrations of organic solvents such as butanol or chloroform/methanol mixtures. After removal of lipids on a Sephadex LH-20 column, the extract was applied on a Synchropak AX 300 anion exchange column. The mobile phase was changed stepwise from the initial chloroform : methanol (2 : 1) mixture to chloroform : methanol (1 : 1), to a final chloroform : methanol : water (3 : 3 : 1) system in which a 3 to 20 mM ammonium acetate gradient was generated; this allowed the isolation of the 8 kd extremely hydrophobic protein (Tandy et al. 1983).

4 Reversed Phase HPLC

It has been observed many times by biochemists that hydrophobic proteins and peptides are retarded more than expected during size exclusion experiments with dextran- or agarose-based media, especially at higher ionic strengths. The most probable explanation for this effect is hydrophobic interaction between the polypeptides and the support. Hydrophobic interaction has not been exploited much in the separation of polypeptides until more hydrophobic supports (e.g., benzylated Sepharose CL 6 B) were introduced that adsorb most polypeptides in the presence of 2 to 4 M salt (Hjerten 1973; Porath et al. 1973). This process is also known as "salting out" of proteins. Desorption from the matrix and elution from the column is simply done by decreasing the salt concentration. Since the number of hydrophobic domains and the degree of their hydrophobicity differs over a wide range between individual proteins and peptides, hydrophobic interaction chromatography (HIC) offers a unique selectivity for polypeptide separation. The high performance counterpart of this technique, HI-HPLC, uses the same mode of separation modified by the characteristics of microparticulate HPLC packings. On these supports, hydrophobic polypeptides are often not desorbed at low ionic strength and require the addition of an organic solvent to the mobile phase for elution. If the density of the hydrophobic ligands on the surface of the hydrophilic silica is increased considerably, polypeptides will be adsorbed even in the absence of salts. The degree of adsorption depends on the hydrophobic character of the ligands. Thus increased percentages of organic solvent in the aqueous mobile phase are needed to displace polypeptides from more hydrophobic ligands. This high performance chromatographic technique, that employs a nonpolar stationary phase and a polar mobile phase with an organic solvent gradient, is known as reversed phase (RP) HPLC. Although the mechanism is not very well understood, experimental evidence indicates that RP-HPLC of polypeptides is an adsorption/desorption process, which in the case of proteins may involve multiple sites on the solute molecule. Since polypeptides differ considerably in hydrophobic properties, and because of the wide variety of reversed phase supports and mobile phase systems that can be used, RP-HPLC offers a tremendous selectivity for the separation of polypeptides. This is clearly reflected in its popularity among protein chemists and the very large number of applications described in literature. In this chapter we will summarize and discuss properties of column packings, mobile phases, and other operational parameters, and how they can be used to advantage in the separation of particular polypeptides.

There is a major difference in the approach to obtain optimal results for chromatography of proteins or peptides. In cases where one has to deal with a wide variety of polypeptides in a single experiment, "average" conditions should be adopted and rechromatography of the fractions under different resolving conditions can be done afterwards. RP-HPLC has proven an extremely powerful tool in the separation of peptides, usually with high recoveries. Although a considerable effort has been made to deal with proteins successfully, chromatography of proteins is often more troublesome, especially in case of very hydrophobic molecules. A special section (4.4.2) is reserved to discuss this problem. In addition, protein recovery varies drastically and the biological activity is generally lost after RP-HPLC. Indeed the hydrophobic adsorbtion and organic solvent-promoted desorbtion usually disrupts the three-dimensional structure of proteins. Since RP-HPLC employs gradient elution, dual-pump systems similar to those described for IEX-HPLC should be used. Ultra-violet detection is usually performed at fixed wavelengths between 205 and 220 (peptides) or 280 nm (proteins). Maintenance suggestions are summarized in Sect. 4.5.

4.1 Columns

RP-HPLC packings are usually microparticulate (3–10 micron) and silica based. Since better mass transfer of the solutes is obtained with 3 or 5 micron particles, these sizes are recommended for analytical experiments. Especially when larger polypeptides are chromatographed, resolution is better and retention decreased on these smaller particle packings (Meek and Rosetti 1981). Use of silica limits the mobile phase pH to below 8.0. At more alkaline pH values, the silica will dissolve. However, The Separations Group (Hesperia, California) recently developed a silica-based Vydac 228 TP packing that is stable up to pH 10. An alkali-stable organic RP packing, TSK phenyl HD-5PW is available from Toya Soda and is based on the TSK 5000 PW size exclusion support.

On silica supports, bonded phase consists of hydrocarbon compounds that are covalently attached to the silica surface by reaction as alkylchlorosilane derivatives. The bonds between the hydrocarbons and the silica are stable under a variety of mobile phase conditions, but nevertheless the hydrocarbons leak off slowly during use, especially at $pH < 2$. Hydrophobic ligands are usually unbranched aliphatic hydrocarbon chains of various lengths. RP packings are then commonly named after the chain length of the ligands: C18, C8, C4, C3, C2, and C1. RP packings with phenyl, diphenyl, or cyanopropyl (CN) bonded phases are also available.

Different RP packings allow the researcher to deal with a wide variety of polypeptides in terms of hydrophobicity and size. Although no well-defined rules for the choice of a particular RP packing are available, fair results will be obtained by adopting the following empirical guidelines. As a rule of thumb, resolution and recovery of large and/or hydrophobic polypeptides are increased by using packings with decreased chain length. C18 packings offer very high selectivities for peptides up to 20 residues in length. Thus complex mixtures of small peptides are best separated on this support as shown in Fig. 2. Most water-soluble larger poly-

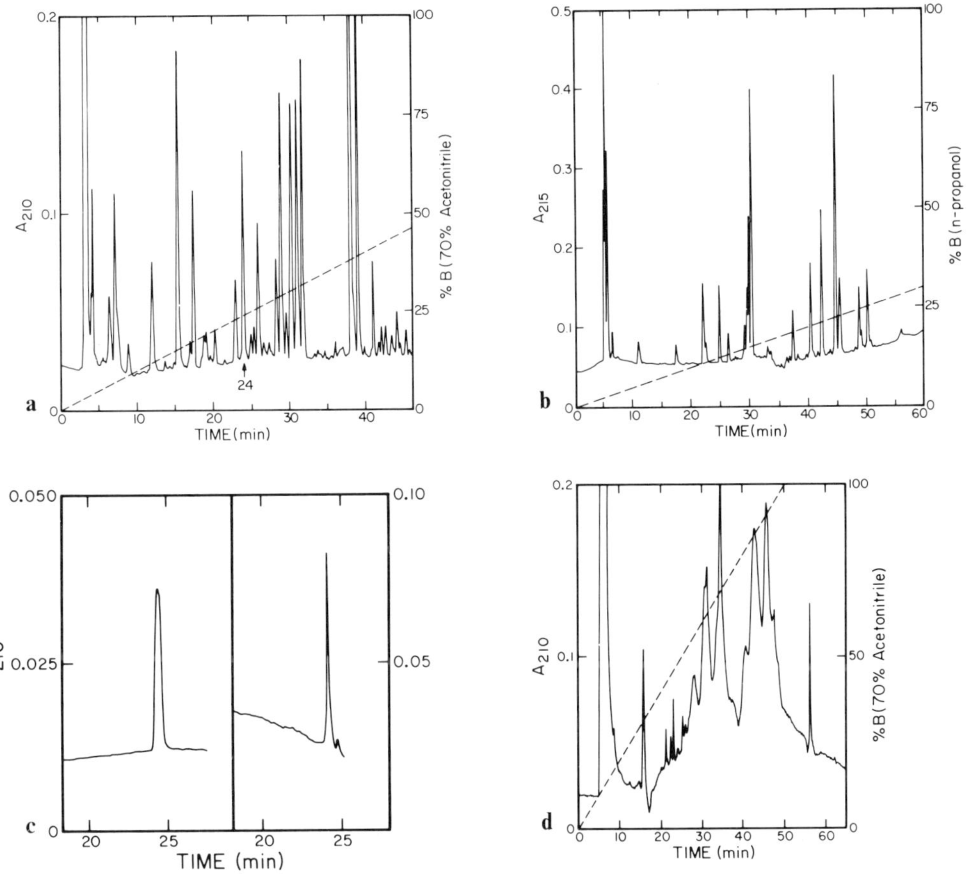

Fig. 1. a Separation of 1.5 nmol horse cytochrome c tryptic peptides (4–15 amino acids) on a 0.46 × 25 cm Vydac 214TP-C4 column (5 micron, 300 Å) with a 0.2 × 2 cm Upchurch C 130 B precolumn hand filled dry with Vydac 214 TP B5 packing. Buffer *A* 0.1% TFA; Buffer *B:* 0.1% TFA in acetonitrile: water (70 : 30). HPLC system: Altex 420 microprocessor controller/programmer, two Altex 110A pumps with a model 110-40 pressure filter, an Altex model 400 mixing chamber, a Rheodyne model 7125 sample injector with a 1 ml loop and a Hitachi model 100-40 variable wavelength detector with a 12 μl flow cell. The sample was dissolved in 5 μl 1% TFA. The flowrate was 1 ml min^{-1}. The system was operated at room temperature and the gradient is indicated as a *dashed line*. **b** conditions as for **a** but buffer *B* was 0.1% TFA in 1-propanol. **c** The peak that eluted after 24 min in **a** was collected and a 50% aliquot injected straight onto the same column under the same conditions (*left hand side*); the other 50% was diluted twofold with buffer *A* before injection (*right hand side*). **d** conditions as for **a** but the peptide mixture was dissolved in 50 μl 0.1% TFA containing 0.5% SDS

peptides will also be eluted from this column but with varying recoveries. C4 and C3 supports offer good resolution for polypeptides ranging in size from five amino acids up to 100 kd (Tempst et al. 1984). The separation of cytochrome C tryptic peptides presented in Fig. 1 a allowed purification and near complete recovery of every peptide larger than four amino acids. Recovery of larger, not too

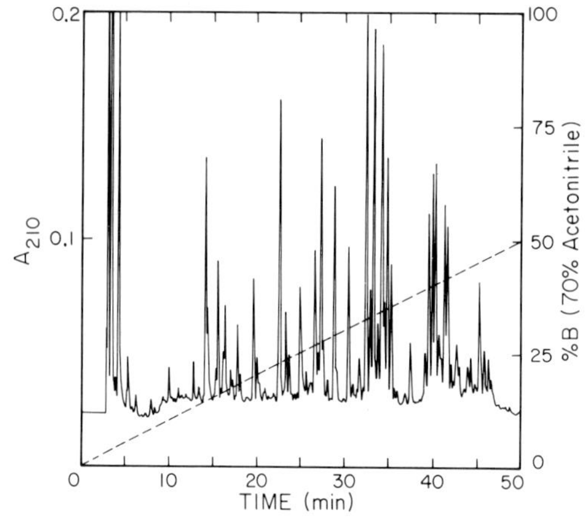

Fig. 2. Separation of 1.5 nmol of horse cytochrome c chymotryptic peptides on a 0.46 × 25 cm Vydac 218 TP-C18 column (5 micron, 300 Å). Conditions as described in the text to Fig. 1 a

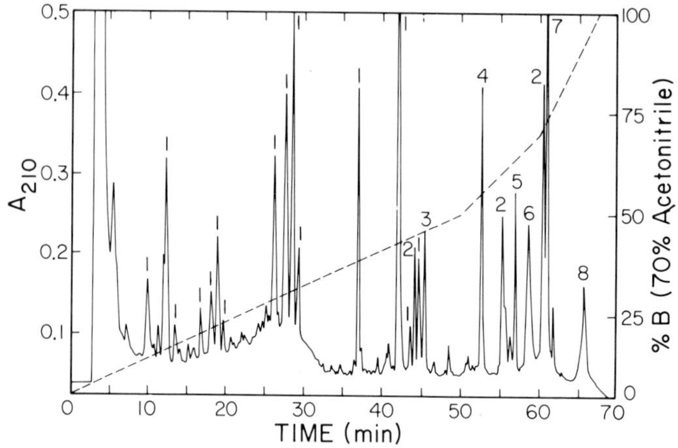

Fig. 3. Separation of a mixture consisting of 3 nmol horse cytochrome c tryptic peptides (*1*), 2 nmol whale myoglobin CNBr fragments (*2*) and 1 μg ribonuclease (*3*), 2 μg cytochrome c (*4*), 1 μg lysozyme (*5*), 3 μ BSA (*6*), 2 μg myoglobin (*7*), and 3 μg ovalbumin (*8*) on a 0.46 × 25 cm Vydac 219 TP-diphenyl (5 micron, 300 Å) column. Conditions were as described for Fig. 1 a except that the injection volumn was 25 μl

hydrophobic polypeptides, such as BSA and ovalbumin, is also fairly good (>85%) under standard conditions (O'Hare et al. 1982). C2 and C1 supports are specifically designed for separations of large and more hydrophobic polypeptides but are not widely used because of limited stability and long-term nonreproducibility. However, successful isolation of hydrophobic viral envelope proteins using a Toya Soda TMS 250-C1 column has been reported (Welling et al. 1984). Phenyl and diphenyl ligands are approximately equal in hydrophobicity to C4 chains, but proteins and medium- to large-sized peptides are slightly more strongly retained, most probably because of phenyl-ring stackings with aromatic residues in the polypeptides. This makes diphenyl packings fairly selective for these polypeptides

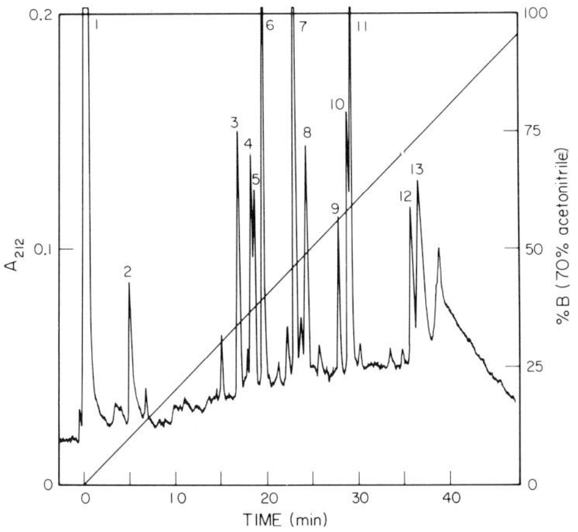

Fig. 4. Separation of 1.9 nmol horse cytochrome c tryptic peptides on a 0.46 × 25 cm IBM cyanopropylsilyl (CN) column. Conditions were as described for Fig. 1 a

as shown in Fig. 3, where medium-sized tryptic peptides, larger CNBr fragments and six proteins are separated in one run. When comparing separations on a C4 column (Fig. 1 a) and a diphenyl column under similar conditions, it is apparent that the latter gives broader peaks and worse resolution for early eluting peptides. Therefore, diphenyl packings are not recommended for smaller peptides. CN packings can be used for proteins and peptides depending on the mobile phase. A separation of tryptic peptides is presented in Fig. 4. Relative elution positions of several peptides differ significantly from those obtained under identical conditions with C4 (Fig. 1 a) and C3 columns (Tempst et al. 1984). Rivier et al. (1982) and Lewis et al. (1980) reported chromatography of larger polypeptides and proteins on a CN column using triethylamine phosphate (TEAP) and pyridineformate buffers. A well-known problem with all CN supports is their limited stability, especially at low pH.

To summarize, different RP packings give significant differences in peptide retention times and resolution, whereas retention times of water-soluble proteins are in the same range although resolution and elution order may vary. Thus, combinations of packings with different types of bonded phases can be used to enhance selectivity in the separation of complex peptide mixtures. Applications of these "two-dimensional" RP-HPLC separations will be discussed in Sect. 4.4.1.

The extent to which silica surfaces are derivatized with a specific hydrocarbon compound is commonly referred to as ligand density or carbonload. There are reports indicating that ligand density is important for resolution and recovery. Low to medium ligand density seems to offer best resolution with C18 packings (Wehr et al. 1982 a; Lindegren et al. 1984) and improved recovery of large polypeptides from C4 columns (Crabb and Heilmeyer 1984). Although the use of different ligand density has not been sufficiently explored, it is clear that it can offer additional selectivity.

The introduction of hydrophobic ligands on the silica surface is never complete, and leaves underivatized silanol groups that are capable of ionic interactions with polypeptide solutes. In order to keep the number of free silanols down, derivatized silicas are additionally reacted with trimethylchlorosilane. This process is commonly known as end-capping. Even under the most vigorous blocking

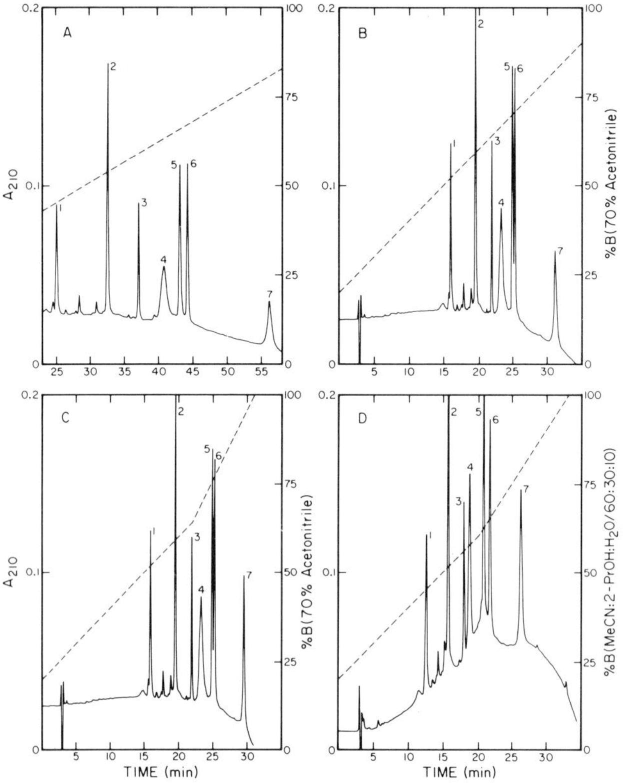

Fig. 5. A–D Separation of a mixture of proteins consisting of 0.75 µg ribonuclease (*1*), 1 µg horse cytochrome c (*2*), 0.5 µg lysozyme (*3*), 1.5 µg BSA (*4*), 1 µg bacterial cytochrome C-556 (*5*), 0.5 µg myoglobin (*6*), and 1.5 µg ovalbumin (*7*) dissolved in 5 µl 0.1%TFA before injection. Column and conditions were as described for Fig. 3; different gradients are indicated by *dashed lines*. **D** As in **A–C**, but buffer *B* was 0.1% TFA in acetonitrile:isopropanol:water (60:30:10)

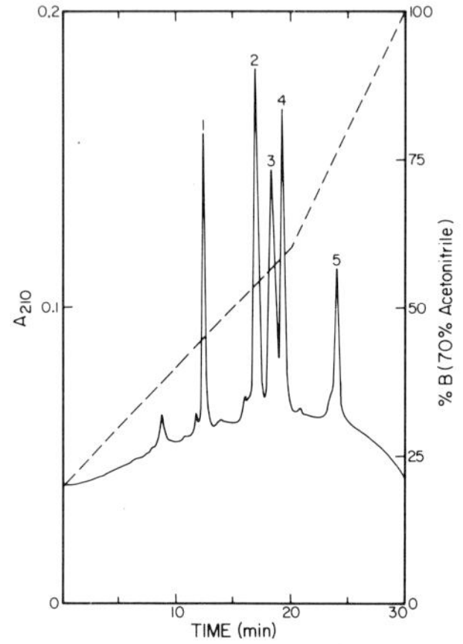

Fig. 6. Separation of 2 µg cytochrome c (*1*), 5 µg BSA (*2*), 2 µg bacterial cytochrome C-556 (*3*), 1 µg myoglobin (*4*), and 5 µg ovalbumin (*5*) on a 2 × 20 mm Upchurch C130 B precolumn handpacked dry with Vydac 214TB5 (C4). Conditions were as described for Fig. 1a

conditions, some free silanols remain at the surface and give the RP packing slight cation exchange properties resulting in mixed mode separations. To avoid this problem, silanol ionization can be repressed by using dilute acids (e.g., H_3PO_4, TFA) in the mobile phase. In addition, oganic amines such as triethylamine (TEA) can be added to compete for the left-over ionized silanols; TEA phosphate buffer at low pH is most effective in this regard.

As pointed out already for IEX-HPLC, the choice of particle pore size depends on the size of polypeptides to be separated. It is now generally believed that 300 Å pore-sized RP packings offer better separations, recoveries and sharper peaks for large polypeptides than those with 80–100 Å pore size (Wilson et al. 1982; Pearson et al. 1981). Smaller peptides are chromatographed with equal success on the large-pore packings (see Figs. 1 a and 2); factors other than pore size are dominant in this case.

Because of the popularity of RP-HPLC in biological sciences, a large number of manufacturers offer a wide variety of RP columns. A literature survey indicates that Vydac (Separations Group), Synchropak (Syn Chrom), Lichrosorb (Merck), Aquapore (Brownlee Labs.), Zorbax (DuPont), and Ultrapore/sphere (Beckman) RP-packings and/or prepacked columns are most commonly used for the separations of polypeptides. Although initial performances of all these packings are approximately of the same level, differences in silica quality and bonding techniques may result in substantial variation in column life times and long-time reproducibility. In addition, variation of ligand density and end-capping efficiencies result in selectivity differences between columns from different manufacturers. Hence, separation of complex mixtures can be maximized by combining such columns in multidimensional RP-HPLC operations (Sect. 4.4.1). It is our experience that VydacTP packings offer good resolution, recovery, and reproduceability for a wide variety of polypeptides. In addition, these column packings are extremely stable and can easily be used for hundreds of hours under acidic conditions without significant deterioration of performance. Our findings are confirmed by a number of colleagues and numerous reports in the literature.

Because RP-HPLC of large polypeptides is primarily an adsorption/desorption rather than a partitioning process, column length mainly affects capacity rather than resolution of protein mixtures. Indeed, it has been reported that decreased column length lowers the capacity for proteins but does not affect resolution in analytical separations (Pearson et al. 1981). To illustrate this fact, we chromatographed 15 µg of a protein mixture on a 0.2×2 cm column handpacked with Vydac C4 large pore support. Although peaks were broader, the result shown in Fig. 6 is comparable to the separation of the same amount of protein mixture on a standard Vydac C4 0.46×25 cm column (data not shown). This, of course, is a very interesting result in terms of purchase price differences. The chromatography of smaller peptides was unsatisfactory on this short column as compared to the separation shown in Fig. 1 a. Whether this negative result was caused by the rather inelegant dry handpacking procedure or by the small column size is not clear at present. However, very good analytical separations of the same peptide mixture were obtained on a 0.4×5 cm C3 Ultrapore RPSC (Beckman) column (Tempst et al. 1984). Unfortunately this column, although five times smaller than regular size columns, does not offer a price advantage; it is also overly sensitive to use and deteriorates during storage under acidic conditions. As a general rule, reduced flow rates improve resolution remarkably when working with short columns (Lindegren et al. 1984).

4.2 Operational Parameters and Resolution

Since RP-HPLC usually employs gradient elution, most theoretical considerations concerning resolution in IEX-HPLC apply as well for this mode of separation. Mobile phase composition is by far the most dominant factor to obtain optimal selectivity in the separation of a particular set of polypeptides. Although determination of optimal conditions is still somewhat an empirical process, understanding of the contribution of different mobile phase characteristics and parameters such as flow rate and gradient slope to the retention and recovery of polypeptides may facilitate the initial choice.

4.2.1 Mobile Phase

There is a popular belief that when polypeptides are adsorbed onto a RP column in an aqueous buffer, they can be desorbed by raising the organic solvent concentration in the mobile phase. This is correct to a certain point. Indeed, there is experimental evidence that the capacity of a silica based RP-support for polypeptide solutes increases dramatically when the organic solvent concentration exceeds 60% (Wehr et al. 1982a; Hearn et al. 1982). As a result, it must also be possible to adsorb polypeptides at 100% organic solvent concentrations and desorb them by gradually decreasing the organic solvent in the mobile phase. Although this offers possibilities for chromatography of organic solvent soluble hydrophobic proteins (Tandy et al. 1983), it has very seldom been used. The bulk of literature about RP-HPLC deals with systems that employ increasing organic solvent gradients and we will focus on this method in the next part of this review. Retention times, elution order and resolution of polypeptides in RP-HPLC are very much dependent on four mobile phase characteristics: (1) choice of organic solvent, (2) mobile phase pH, (3) selection of ion-pairing agent, and (4) acid concentration (in addition to the effect on pH).

4.2.1.1 Organic Solvent

Some important criteria in the choice of organic modifyers in RP-HPLC are the solubility of the polypeptides in the organic solvent/water mixtures, the lack of reactions with the solutes and chemical and physical compatibility with the column supports. There is a significant difference between suitable organic solvents in terms of polypeptide displacing power also known as elutropic strength. Some of the organic solvents commonly used in RP-HPLC of polypeptides are listed below in order of decreasing elutropic strength, this means retention of polypeptides will be increased in the order: 1-propanol (shortest retention times) $>$ 2-propanol (iso propanol) $>$ ethanol $\geqq$ acetonitrile $>$ methanol (Mahoney and Hermodson 1980). From this series, acetonitrile and isopropanol are by far the most popular solvents among researchers today. There are also a few reports about the use of acetone, tetrahydrofurane, and dioxane as mobile phase modifying agents. However, aqueous solutions of methanol and acetone are not particularly good for polypeptide solubilization, hence recoveries are usually low. In the case of hydrophobic polypeptides, methoxyethanol and 1-butanol have also been used in combination with other alcohols (Sect. 4.4.2).

As a general approximation, acetonitrile is the solvent of choice for the separation of peptides, while isopropanol is preferred when larger polypeptides are chromatographed (Wilson et al. 1981 b; O'Hare et al. 1982). Indeed, as can be seen in Fig. 1 a and b, acetonitrile gradients give better separations of smaller peptides than propanol. Acetonitrile is an excellent solvent for use in RP-HPLC of polypeptides in general because it is UV-transparent, has low viscosity, and polypeptides are fairly soluble in acetonitrile/water mixtures. However, it has been observed that acetonitrile gradients sometimes cannot elute large denaturated polypeptide fragments from RP supports, whereas the use of propanol was successful in doing so (Mahoney and Hermodson 1980). Since propanol is a stronger elutropic solvent, large polypeptides are indeed eluted faster and with better recoveries. As a general rule polypeptide recoveries from RP supports decrease with increasing organic solvent concentrations needed for elution (Kerlavage et al. 1983). This may be partially due to the reduced protein solubility in higher concentrations of organic solvents. The disadvantage of propanol is its higher viscosity than either acetonitrile or water. As a result of this, column back pressure increases strongly during the experiments, a substantial baseline drift occurs (depending on detector flow cell design) and peak widths are considerably broader than with acetonitrile. Several groups, however, have reported the use of acetonitrile/propanol mixed organic solvents for the chromatography of proteins with results that exceed the performance of either solvent alone. We chromatographed seven standard proteins on a Vydac diphenyl, large pore support with acetonitrile (Fig. 5 c), isopropanol (data not shown) and different mixtures of the two as modifying solvents. Best resolution and peak shapes were obtained with an acetonitrile : isopropanol (2 : 1) mixture; the result is shown in Fig. 5 d.

Since the capacity of a RP support for polypeptides is at its lowest point between 50 and 70% organic solvent in the mobile phase, there is no need to increase the concentration to 100% to accomplish elution of strongly retained polypeptides. This will only result in a tighter adsorbtion to the support. Instead, alternative elution procedures, as discussed in Sect. 4.4.2, have to be explored. We routinely use 70% organic solvent in buffer B of our systems. This also allows thorough mixing and degassing by sonication (2 min) of the buffer with concomittant reduction of possible bubble formation and/or baseline drift due to refraction index changes during generation of the gradient.

4.2.1.2 Mobile Phase pH

The mobile phase pH affects the behavior of polypeptide solutes on a RP column by two principal mechanisms. Because of the stationary phase stability restrictions, RP columns are usually operated in the pH range 2 to 7.5. When the pH is decreased from neutral to acidic, protonation of α-amino groups and basic amino acid side chains occurs and peptides become less polar. This results in increased retention of small neutral or basic peptides, but hydrophobic peptides and proteins in general are not affected. Experimental evidence indicates that small polar peptides are indeed well separated under acidic conditions, whereas they coelute or elute clustered together early in the separation when a neutral pH mobile phase is used (Rivier 1978; Wilson et al. 1981 a, 1982; Kratzin et al. 1982, 1984). In addition, ionization of free surface silanols is repressed at low pH and undesirable ionic interactions between the RP support and polypeptide are thus eliminated. This is reflected in a sharper polypeptide peak shape at low pH as compared to neutral pH conditions. Dilute concentrations ($\sim$0.1%) of mineral acids, e.g., H_3PO_4 (Fullmer and Wasserman 1979) or organic perfluorinated carboxylic acids, e.g., trifluoroacetic acid (Mahoney and Hermodson 1980) are the most simple and common ways to assure low pH of the mobile phase. Alternatively, dilute H_3PO_4 is sometimes titrated to pH 2.25 to 3.5 with triethylamine (TEA). TEA competes for negatively charged ionized silanol groups thus eliminating aspecific support-solute interactions even more than under acidic conditions only. The latter type buffer system is commonly referred to as TEAP and has been used with success (Rivier 1978; Rivier et al. 1982). It is also effective at neutral pH and offers additional versatility in RP-HPLC elution conditions.

4.2.1.3 Ion-Pairing Agents

Depending on pH, specific side groups of basic and acidic residues of polypeptides in solution are ionized and carry positive or negative charges respectively. Appropriate counterions added to the solution can thus form ion pairs with the polypeptides and the resulting complexes with no net charge will interact as a whole with the stationary phase during RP-HPLC. In this way, depending on the nature of the ion-pairing agent (IPA), the hydrophobic characteristics of the polypeptide-IPA complexes can be altered. This offers a most useful tool to change the selectivity of RP separations. Since most separations are performed under low pH mobile phase conditions, the extent to which polypeptide retention will be affected by the choice of IPA depends on the number of basic amino acids (Lys, Arg, His) and suitable IPA's should of course be anionic (negatively charged).

A first class of IPA's that can be used for this purpose are polar mineral-acid anions such as $H_2PO_4^-$, HPO_4^{2-}, SO_4^{2-}, and ClO_4^-. TEAP buffers and 0.1% H_3PO_4 are examples of mobile phases that offer these hydrophilic ion-pairing capabilities. $NaClO_4$, when used as mobile phase additive at low pH, is not only a hydrophilic IPA but also acts as a chaotropic agent that disrupts polypeptide structure and limits conformational effects during RP-HPLC. A mobile phase consisting of 0.1 M $NaClO_4$ in 0.1% H_3PO_4 (Meek 1980) will also result in decreased retention times for most polypeptides as compared to IPA systems discussed below because of the relatively polar nature of ClO_4^-. Its use can be an advantage in the separation of rather hydrophobic polypeptides. For instance, better separations of 1 M urea insoluble bone proteins could be obtained with this system than with 0.1% TFA (Tempst, unpublished observations). A second class of common IPA's in RP-HPLC are organic perfluorinated carboxylic acids. Trifluoro acetic acid (TFA) is most popular in this regard. It is volatile, UV-transparent and a good solubilizing and denaturing agent for polypeptides. Ion pair formation of TFA with polypeptides will render the latter more hydrophobic and increase the retention time on RP supports. Retention times of peptides on a CN support using either TFA or $H_3PO_4/NaClO_4$ as IPA's are compared in Table 4. Aqueous 0.1% TFA/acetonitrile as mobile phase offers very good results in the separations of peptides with any RP support (Figs. 1a and 2–4) but can also be used with success for chromatography of proteins (Fig. 5). Alternatively pentafluoropropanoic acid (PFPA), heptafluorobutyric acid (HFBA) and undecafluorocaproic acid (UFCA) have been used as hydrophobic IPA's in RP-HPLC of polypeptides (Bennett et al. 1980). Hydrophobicity and retention of the complexes is increased proportionally to the fluorocarbon chain length of these organic acids and also depends on the number of basic amino acids of the polypeptides. Substantial differences in resolution and elution order of peptides can be obtained by using different perfluorinated carboxylic acids as additives to the mobile phase. Alkylsulphonates with 4 up to 12 carbon atoms make a class of very hydrophobic IPA's, and have been used in the separation of small very polar peptides. Spindel and Wurtman (1979) separated thyrotropin-releasing hormone (Glu-His-Pro) analogs on a C18 column with an acetic acid/acetonitrile mobile phase supplemented with 0.1% 1-heptanesulphonic acid. When a column is operated at moderately low pH, some of the acidic residue side chain groups may not be protonated and will still carry a negative change. Positive counterions can then form ion pairs. There is a possibility that TEA in buffers of pH 2.5–3.5 has an additional effect on polypeptide retention in this regard.

4.2.1.4 Other Aqueous Solvent Compositions

Besides the mobile phase systems described above, a number of other aqueous solvent compositions for RP-HPLC have been described in literature. NaCl (0.15 M) at low pH, 10–100 mM sodium or potassium phosphate and ammonium acetate buffers at pH 4 to 7, and 50 mM Tris chloride pH 7.4 have all been used for the separation of peptides. Concentrated pyridine formate or acetate buffers (0.5–1.5 M) were used for chromatography of large polypeptides, sometimes in combination with fluorescent detection (Lewis et al. 1980). An excellent overview of different mobile phase systems and their practical application is given by Hearn (1984).

4.2.1.5 Acid Concentration

It has been pointed out by Geng and Regnier (1984) that, whatever organic solvent used, retention of polypeptides on RP supports will decrease with increasing acid concentrations in the mobile phase. Retention times of three different peptides chromatographed on a CN-column using TFA concentrations varying from 0.01 to 0.2% TFA clearly illustrate this observation (Table 4). Addition of relative high concentrations (5–60%) of formic acid to the mobile phase will result in elution of most peptides with the column void volume and in a much decreased retention of polypeptides that are strongly retained or not eluted at all under normal conditions. Also because aqueous formic acid is a very good solubilizing

Table 4. Effect of ion pairing agent and acid concentration on RP-HPLC peptide retention

% TFA in mobile phase	Elution time (min)		
	Peptide 2	Peptide 3	Peptide 13
0.01	18	N.E.	N.E.
0.06	10	22	47
0.08	8	20	39
0.1	6.5	18	33
0.2	5	16	24
0.1 M $NaClO_4$ in 0.1% H_3PO_4	4	11	23

Chromatography was performed on an 0.46 × 25 cm CN-column (IBM) under conditions as described in the legend of Fig. 4. Peptides correspond to the numbered peaks in Fig. 4; their sequences are as follows: 2, GDVEK; 3, YIPGTK; 13, TGPNLHGLFGR, N.E. means not eluted.

agent for proteins, addition to the mobile phase has allowed RP-chromatography of hydrophobic proteins. Examples are described in Sect. 4.4.2 of this review.

4.2.2 Mobile Phase Velocity and Temperature

In analogy to IEX-HPLC, decreased flow rates will increase resolution of proteins and peptides in RP-HPLC experiments, especially when high molecular weight polypeptides are involved (Meek and Rosetti 1981). For reasons of pump accuracy, the lower limit with most gradient HPLC systems is 0.25 ml min^{-1}. For most routine separations on a 0.46 × 25 cm column, a standard flowrate of 1 ml min^{-1} is usually employed because slightly increased peak heights and sensitivity (Tempst et al. 1984) and better recoveries of hydrophobic polypeptides (O'Hare et al. 1982) are obtained than with the better resolving, decreased flowrates. In case of very complex separations as, for instance, tryptic peptides of fairly large proteins, a flow rate of 0.65 ml min^{-1} gives a good compromise between resolution, sensitivity, and recovery (see Fig. 7b). As a general rule, increased temperature will not affect the separation of polypeptides measured at the peak centers, but will give sharper peaks, thus increasing resolution. However, an efficient combination of other operational parameters can often make the use of elevated temperatures unnecessary.

4.2.3 Gradient Slope

As a rule, better separations of polypeptides are obtained when gradient slopes are decreased. For example, separation of highly similar insulins from different animals was obtained using a C4 column with a very shallow 45 to 50% acetonitrile gradient over 25 min in 0.1% TFA (Rivier and McClintock 1983). However, peak shape and recovery of some more retained proteins can be much improved by a steeper gradient (O'Hare et al. 1982). The influence of gradient slope on the resolution of seven proteins and the peak shapes of BSA and ovalbumin, chromatographed on a diphenyl support, is shown in Figs. 5a–c. For all separations in which larger polypeptides are involved, we routinely increase the gradient slope towards the end of the run in one or two steps can be seen in Figs. 3, 6, and 7b. On the other hand, very polar peptides sometimes need an isocratic prewash of 5 to 10 min with the starting buffer to improve resolution.

4.2.4 Sample Composition and Load

As for other types of HPLC, viscosity of the sample can affect resolution of RP chromatography and should be kept low. In addition, certain sample additives also affect resolution in a very dramatic way. We found that the addition of 100 μl of either 4 M urea, 2 M guanidine · HCl or 1% Triton X-100 to a 5 μl peptide mixture sample did not affect the peptide separation as presented in Fig. 1a. However, when 50 μl of 0.5% SDS was added, we obtained the disastrous result shown in Fig. 1d. This has been generally observed by researchers, and although the exact cause is not known, ion-pairing with the polypeptides and hydrophobic interaction with the RP support are most probably involved. High concentrations of TFA or formic acid are sometimes used to solubilize polypeptides and 70% formic acid is also a common solvent for CNBr digests. Since acid concentration in the mobile phase affects polypeptide retention, sufficiently large volumes of samples dissolved in these concentrated acids will also influence peptide chromatography. Addition of 500 μl 70% formic acid to the sample completely ruined the separation, shown under normal conditions in Fig. 1a; 100 μl 70% formic acid caused all the peaks that are retained 5 to 15 min under standard conditions to elute with the column void volume. The overall effect of sample acid concentration on protein retention was found to be minimal. When the mobile phase contains 0.1% TFA, acidification of the sample can improve peptide separations especially when salts are present. Therefore, we usually stop enzymatic digests by addition of TFA to final concentrations of 1–5% prior to RP-HPLC. It is easy to understand that polypeptides will not adsorb to a RP support instantly when sufficiently high concentrations of organic solvent are present in the sample. When fractions collected from RP-HPLC are to be rechromatographed on another RP column, they should be diluted at least twofold with buffer A prior to injection. Failure to do so results in broad peaks (Fig. 1c) and decreased resolution.

Provided that sample composition is properly adjusted, polypeptides will be adsorbed tightly to RP supports in a mobile phase free of organic solvent. In this way, large volumes can be applied to the column resulting in concentration of dilute sample solutions. When the peptide mixture described in the legend to Fig. 1a was diluted with 0.1% TFA from 5 μl to 2 ml, applied to the C4 column in multiple injections while buffer A was pumped over the column and then chromatographed, a separation identical to that shown in Fig. 1a was obtained. When a standard column (0.46 × 25 cm) was used, protein loads ranging from a few micrograms to 30 mg could be applied (Pearson et al. 1982) and subnanomole to approximately 50 nmol amounts of peptide mixtures can be separated with good resolution. When larger quantities need to be chromatographed as for instance in the purification of genetically engineered proteins or synthetic peptides, column size should be increased. Excellent studies in this regard were done by Pearson et al. (1982). Rivier et al. (1984) and Crabb and Heilmeyer (1984). Basically, column volumes should be proportional to the load. Scaling-up of columns is best done by increasing the inner diameter; length is less important. Flow rates are then also increased in proportion to the increased cross-sectional area. Low ligand density packings are most useful for preparative purposes and the less expensive 17 to 30 micron particle supports offer sufficient resolution.

4.2.5 Prediction of Retention Times

The availability of different RP supports and an imaginative choice of operational conditions offer almost unlimited selectivity in RP-HPLC and allow the researcher to deal with a wide variety of polypeptide mixtures. Since there exist fair correlations between most operational parameters and polypeptide retention on a specific RP-column, it is often possible to predict retention times. Reports from Meek (1980), Meek and Rosetti (1981), Su et al. (1981), and Sasagawa et al. (1982) indicate that retention contributions of the individual amino acids have an essentially additive effect on peptide retention with RP supports. However, absolute retention predictions are only valid within the limits of one particular set of operational conditions and predicted elution times for peptides bigger than 15–20 amino acids are usually inaccurate. For the moment, predictions of protein retention are difficult, if not impossible, and there is no definite correlation between molecular weights and retention (see Fig. 5).

4.3 High Sensitivity RP-HPLC

Following the introduction (Hewick et al. 1981) and commercial availability of the gas-phase sequenator, amino acid sequence analysis at the low (5–20) pmol level came within reach of a number of protein research laboratories. Chromatography of a few hundred nanograms of most proteins (1–20 pmol, depending on size) can be monitored with conventional UV detectors but, unfortunately, the isolation of smaller polypeptides by RP-HPLC at this level has lagged behind. The lower limit of sensitivity in practical UV flow detection of peptides eluted from standard 4.6 mm columns is barely under 100 pmol. Stream-splitting and postcolumn derivatization with subsequent fluorescent detection has been used but in practice does not increase detection sensitivity significantly. Precolumn derivatization with the chromophore dimethylaminoazobenzene isothiocyanate (DABITC) enables the peptide derivative to be detected in the visible region at 5 to 10 pmol levels (Chang 1981) and is compatible with sequence determination of the collected peptides. Traditionally, the separations of DABITC-modified peptides have been carried out at neutral pH with detection at 436 nm; however, chromatography using the popular TFA/acetonitrile solvent system with high-sensitive detection at 510 nm has also been used with excellent results (Tempst, Aebersold and Kent, unpublished observations). The applications of the DABITC-technique reported in the literature describe enzymatic digestion and DABITC-derivatization of nanomolar polypeptide amounts followed by RP-HPLC of low picomolar aliquots. This, of course, does not represent true high sensitive protein structure analysis and more work is needed to successfully adapt these derivation techniques to protein quantities at the <50 pmol level.

A recent and very promising approach to high-sensitivity HPLC is microbore RP-HPLC. For this purpose, narrow (1–2 mm) and short columns (10–50 mm) are used in combination with reduced flow rates (50 µl – 0.2 ml min^{-1}). This limits the dilution of the solutes during the run and results in smaller peak fraction volumes, substantially increased polypeptide concentrations and consequent increased sensitivity. However, standard HPLC gradient instrumentation is usually inadequate for these operations and new, more sophisticated systems should be designed (Pearson 1984). Some of the best low flow rate gradient HPLC systems that are currently available use microprocessor-controlled syringe pumps and are manufactured by Brownlee Labs (Santa Clara, California). If only standard HPLC equipment is to be used with a semi-microbore column, adequate gradient elution can be performed by applying the polypeptides to the column in buffer A at normal flow rates (0.2–1 ml min^{-1}), then decreasing the flow to 50–200 µl min^{-1} and simultaneously switching the mobile phase composition to that desired at the end of the gradient. The gradient will then be formed in the dead volume of the mixing chamber. Separations and standard UV detection of peptide mixtures containing as little as 10 pmol of each component can be obtained this way (Nice et al. 1984; Hunkapiller et al. 1984).

4.4 Applications

4.4.1 Two-Dimensional RP-HPLC

The resolution of components of complex protein or peptide mixtures can be greatly improved by rechromatography of collected fractions using different RP conditions. One can either keep the same column but change one or more characteristics of the mobile phase, chromatograph on a column support with the same type of hydrophobic ligand but from a different manufacturer or use a RP packing with a different ligand type. Each of these three versions of two-dimensional RP-HPLC has been described in the literature.

Isolation of corticotropin (39 amino acids) and related peptides from pituitary extracts was accomplished by chromatography on a C18 column with an acetonitrile gradient in 0.1% TFA; the collected peaks were further fractionated using the same column and organic modifyer but with 0.13% HFBA in the mobile phase (Bennett et al. 1981). Enzymatic cleavage products from human histocompatibility antigens, integral membrane proteins, were initially separated on a C18 column with 50 mM ammonium acetate pH 6.0/acetonitrile as mobile phase. Early eluting peptide fractions were further purified on an identical column operated with 0.1% TFA/acetonitrile and the remaining fractions were fractionated using a 5 mM potassium phosphate pH 6.0/acetonitrile system (Kratzin et al. 1984). A somatostatin-like peptide (14 amino acids) from fish pancreas was purified for sequencing studies by Andrews and Dixon (1981) in a two-step RP-HPLC procedure that involved consecutive runs on a Lichrosorb RP-18 and a Vydac C18 (300 Å pore size) column, both eluted with 0.1% TFA/isopropanol. Berchtold et al. (1982) used Lichroprep RP-18 and Lichrosorb RP-18 columns in combination with a 50 mM Tris chloride pH 7.5/isopropanol elution system followed by two passages over an Aquapore RP 300 column using a pH 7.5 Tris · HCl/acetonitrile and 20 mM ammonium acetate/acetonitrile as mobile phases in the complete purification of rat neuronal parvalbumin (12 kd).

For high resolution, two-dimensional separations of complex peptide mixtures on different types of RP columns, it is recommended to start with a column that allows fair separation of a wide variety of polypeptides such as C4 or C3 supports. Early eluting small and/or polar peptides can then be rechromatographed with different selectivity on C18 or CN columns; later eluting larger or more hydrophobic polypeptides are best further separated on diphenyl, C2 or C1 columns. We have applied this in our laboratory for the separation of platelet-derived growth factor receptor tryptic peptides. More than 40 peptides were isolated from approximately 250 pmol digest mixture using Vydac C4, C18 and diphenyl columns, all operated in 0.1% TFA/acetonitrile. Some of these separations are shown in Figs. 7b and 7c (A–D).

4.4.2 Hydrophobic Polypeptides

Because of the mode of interaction with the stationary phase in RP-HPLC, hydrophobic polypeptides such as membrane proteins and fragments thereof are strongly retained, poorly recovered, and sometimes not eluted at all. Improved separations and recoveries may be obtained by treating the polypeptides with anhydrous TFA at room temperature prior to chromatography. After evaporation of the TFA under a stream of N_2, samples can be dissolved in suitable solvents such as dilute TFA or formic acid or solutions of guanidine hydrochloride or urea, and injected into the HPLC. When proteins are not eluted under standard conditions, one can try to recover them by injection of 1 ml DMSO on the column operated isocratically at 60% propanol. However, best results in RP-chromatography of large hydrophobic polypeptides have been obtained with modified solvent systems in combination with large pore supports. An acetonitrile : propanol (2 : 1) mixture as organic modifyer not only gives increased resolution but also dramatically improved recoveries of membrane proteins such as cytochrome P-450 and rhodopsin (Tarr and Crabb 1983). The use of butanol/ethanol mixtures and addition of 10% 2-methoxyethanol to both buffers were also reported to be beneficial in this regard (Van Der Zee and Welling 1982).

As pointed out before (Sect. 4.2.1.5), increased acid concentrations in the mobile phase will increase the ability of organic solvents to displace polypeptides from RP supports. Elevated concentrations of TFA and formic acid have been used to advantage in the separation of large and/or very hydrophobic polypeptides. The different hemoglobin chains could be separated on a Synchropak RP-P column with an acetonitrile gradient in 0.5% TFA (Mahoney 1982). Gerber et al. (1979) were able to separate CNBr fragments of bacteriorhodopsin, an integral membrane protein, on a Microbondapak C18 column (0.4×30 cm) with a linear ethanol gradient in 5% formic acid at a flow rate of 2 ml min^{-1}. Formic acid concentrations of 60% in the mobile phase keep almost any protein solubilized and allow elution of the most hydrophobic protein with less than 30% propanol (Heukeshoven and Dernick 1982). The authors applied this system with success to the separation of the four structural proteins of poliovirus on an Aquapore RP-300 column whereas those proteins could not be eluted with any of the mobile phase systems described higher. Similarly, Kratzin et al. (1984) used C18 Sep-pak cartridges (Waters Assoc.) to separate hydrophobic from hydrophilic membrane protein fragments by adsorbing the complete peptide mixture onto the cartridge in a phosphate buffer followed by elution with stepwise increased propanol concentrations in 60% formic acid. All polar peptides eluted between 0 and 5% propanol while an intramembranous chymotryptic fragment was eluted with 25% propanol. Alternatively, the pH of moderately concentrated formic acid solutions can be adjusted by adding organic bases to the mobile phase. Tarr and Crabb (1983) chromatographed denatured membrane proteins as sharp peaks on different RP-columns employing a gradient from 6 M formic acid/0.13 M trimethylamine (pH 1.5) to 4 M formic acid/0.09 M trimethylamine in 72% 1-propanol. Johnsson et al. (1982) separated the two polypeptide components of platelet-derived growth factor (PDGF), a very adhesive molecule, on a Lichrosorb RP-8 column. To achieve this, reduced and alkylated PDGF was dissolved in 1 M acetic acid containing 1 mg ml^{-1} gelatin as a carrier, injected onto the column and eluted with a 0–26% propanol gradient in 2 M formic acid/0.8 M pyridine at 0.4 ml min^{-1} over 3 h. The authors also used a Synchropak RP-P (C8) column operated in 2 M guanidine·HCl/1 M acetic acid for the same purpose (Johnsson et al. 1984). Chromatography with high concentrations of formic acid limits the UV absorbance detection to 260 nm and above and prolonged drying of the collected fractions may cause formylesterification of the Ser and Thr hydroxyl groups. Esterification is also possible during or after chemical cleavages in aqueous formic acid (e.g., CNBr or Asp-Pro cleavage) but can be reversed by treatment with neat aminoethanol (Tarr and Crabb 1983). Failure to do so results in unusually increased retention and peakbroadening during eventual subsequent RP-HPLC.

4.4.3 Others

The RP separations discussed below deserve careful examination because they deal with rather complex protein mixtures. The *E. coli* ribosome consists of one large and one small subunit that contain 33 and 21 proteins respectively with a wide variety in molecular weights (5–29 kd and one protein of 61 kd) but mostly basic. Good separations of these proteins were obtained using a Synchropak RP-P (C18) column with a 0.1% TFA/acetonitrile mobile phase and a flow rate of 0.7 ml min^{-1} (Kerlavage et al. 1983) or a Vydac C18 (300 Å pore size) column operated in 0.1% TPA/isopropanol (Kamp et al. 1984). In both experiments gradients were very shallow and sometimes interrupted by short periods of isocratic elution. The high resolving power of RP-HPLC under "standard conditions" was also demonstrated by Bietz (1983) in the separation of cereal proteins on a large pore (300 Å) Synchropak C18 support, 0.1% TFA as ion-pairing agent and acetonitrile as organic modifyer. Although these proteins are known to be very heterogenous, poorly soluble and to polymerize easily, wheat low molecular weight gliadin was resolved into 36 peaks and omega-gliadin into 25 components; 70% ethanol or 50% acetic acid extracts from single kernels of a wheat variant were resolved into more than 30 clearly separated protein peaks.

4.5 Maintenance

To avoid adsorbtion to the top of the RP column of any particulate material or irreversible bonding and/or precipitation of polypeptides, a guard column should always be used. This also helps minimize dissolution of silica from the column when operating at higher pH values. We found that cheap, self-packed guard columns perform as well as the rather expensive prepacked ones. The top frit and the guard column should e regularly checked for particle buildup. In case proteins are irreversibly adsorbed or precipated on the main column one can try to clean it by, for instance, injecting DMSO and washing the column with 60% propanol or heating the column at 60 °C in 70% acetic acid saturated with guanidine · HCl followed by propanol gradient elution. Column operation at low pH may cause the bonded phase to leak off. However, we used Vydac RP columns for more than 200 h of operation and more than 6 months of storage in 0.1% TFA without significant change of performance. When 60% formic acid is used in the mobile phase, the complete HPLC system and column should be rinsed very thoroughly with water after the experiments and should be kept in water overnight in case of long-term use. Failure to do so will result in severe column damage and corrosion of all metal parts. Long term storage of RP-HPLC columns is preferentially done in organic solvents.

5 Multidimensional HPLC

For several complex polypeptide separations, optimal selectivity can be obtained by a combination of any two or all three of the above described chromatographic techniques. It is important that a polypeptide solution collected as a column fraction, in order to be chromatographed well on the next support, meets optimal sample characteristics as discussed in the previous sections. For practical reasons, the sequence of most common combinations of HPLC techniques is as follows:

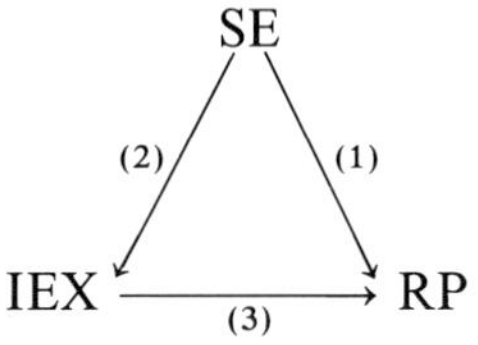

1. Remove SDS, dilute out organic solvent and acidify when appropriate.
2. Remove salt or decrease ionic strength and adjust pH when appropriate.
3. Acidify when appropriate.

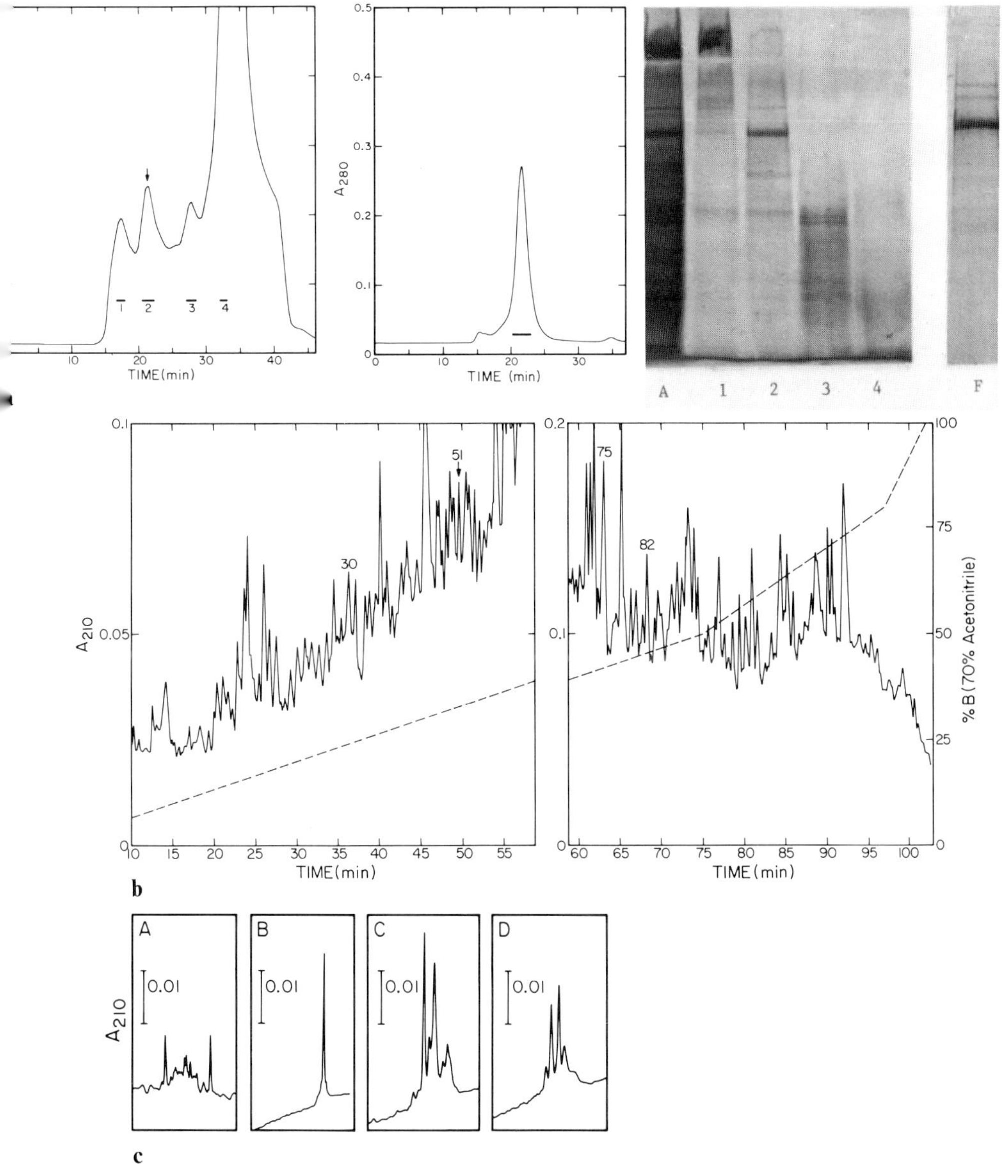

Fig. 7. a Chromatography of partially purified mouse PDGF-receptor on a 0.75 × 30 cm TSK G 4000 SW column eluted with 0.1% SDS in 0.1 M sodium phosphate pH 7.0 (*far left*). Collected fractions are indicated with *heavy bars* and 1% was analyzed on SDS-PAGE (*far right*). Fraction 2 was concentrated and rechromatographed under the same conditions (*middle*), and analyzed by SDS-PAGE (*far right-fraction F*). Lane *A* contains the initial affinity purified (pre SE-HPLC) material. **b** Separation of PDGF-receptor tryptic peptides on a Vydac 214 TP (C4) column and C4 precolumn. Conditions were as described for Fig. 1a but the sample was applied in 600 µl 2 M urea/50 mM NH_4HCO_3 and a flowrate of 0.65 ml min^{-1} was used. **c** Fractions 30, 51, 75, and 82 collected in **b** were diluted twofold with buffer A and rechromatographed on Vydac C 18 *A*, Vydac C4 *B*, and Vydac diphenyl columns *C*, *D*, respectively, under conditions as described in Figs. 1a, 2, and 3

RP-HPLC is best reserved to the final step since it offers excellent resolution, sharp peaks and small collection volumes and can be easily done with completely volatile mobile phases. SE- and IEX-HPLC give good initial group separations. Multidimentional HPLC has been used for isolation of proteins, peptides or a combination of the two.

RP-HPLC of peptide fragments of pro-opiomelanocortin was facilitated by prior separation into a basic, acidic and neutral pool using cation- and anion-exchane columns (James and Bennett 1985). A fairly similar but automated version was used by Takahashi et al. (1985) to fractionate a ceruloplasmin tryptic digest into more than 300 peaks. Ionic strength gradient-elution of an AEX column was done stepwise and collected eluate corresponding to two column volumes was periodically introduced directly onto a RP column which was developed with a linear organic solvent gradient. Stepwise and alternate elution of the AEX and RP column was synchronized by a computer program.

Finally, we would like to describe briefly the purification of the PDGF-receptor, a 185 kd intrinsic membrane glycoprotein, and the separation of its tryptic peptides using HPLC techniques only. The protein was purified from Triton X-100 solubilized mouse fibroblasts by medium performance affinity chromatography on a TSK fractogel-ricin B column followed by SEC on TSK 4000 SW column, in the presence of SDS (Fig. 7a). Parallel analytical SE runs were done in the presence of Triton X-100 to monitor the biological activity. Because of the extreme insolubility of the purified receptor, methods to produce fragments were developed which did not involve precipitating or drying the protein. Thus, the PDGF receptor was alkylated in 8 M urea, dialyzed against decreasing concentrations of urea, cleaved with trypsin in 2 M cyanate free urea, immediately injected on a Vydac C4 column and eluted with an acetonitrile gradient in 0.1% TFA. The resulting separation is shown in Fig. 7b. Peptides were further purified by rechromatography on Vydac C4, C18 and diphenyl columns using the same gradient system. Four examples of rechromatography are presented in Fig. 7c (A–D). Starting with 50–60 µg (250–300 pmol) of purified receptor, 25 peptides were isolated and sequenced, yielding over 240 residues of sequence.

References

Alfredson TV, Wehr CT, Tallman L, Klink F (1982) Evaluation of new microparticulate packings for aqueous steric exclusion chromatography. J Liq Chromatogr 5:489–524

Andrews PC, Dixon JE (1981) Isolation and structure of a peptide hormone predicted from a mRNA sequence. J Biol Chem 256:8267–8270

Barth HG (1982) High-performance size exclusion chromatography of hydrolyzed plant proteins. Anal Biochem 124:191–200

Bennett HPJ, Browne CA, Solomon S (1980) The use of perfluorinated carboxylic acids in the reversed-phase HPLC of peptides. J Liq Chromatogr 3:1353–1365

Bennett HPJ, Browne CA, Solomon S (1981) Biosynthesis of phosphorylated forms of corticotropin-related peptides. Proc Natl Acad Sci USA 78:4713–4717

Berchtold MW, Wilson KJ, Heinzmann CW (1982) Isolation of neuronal parvalbumin by high performance liquid chromatography. Biochemistry 21:6552–6557

Bietz JA (1983) Separation of cereal proteins by reversed phase high performance liquid chromatography. J Chromatogr 255:219–328
Chang J-Y (1981) Isolation and characterization of polypeptide at the picomole level. Biochem J 199:537–545
Clezardin P, McGregor JL, Manach M, Robert F, Dechavenne M, Clemetson KJ (1984) Isolation of thrombospondin released from thrombin-stimulated human platelets by fast protein liquid chromatography on an anion-exchange mono-Q column. J Chromatogr 296:249–256
Crabb JW, Heilmeyer LMG (1984) Micropreparative protein purification by reversed phase high performance liquid chromatography. J Chromatogr 296:129–141
Delucas LJ, Muccio DD (1984) Purification of bovine rhodopsin by high performance size exclusion chromatography. J Chromatogr 296:121–128
Downward J, Yarden Y, Mayes E, Scrace G, Totty N, Stockwell P, Ullrich A, Schlessinger, Waterfield MD (1984) Close similarity of epidermal growth factor receptor and v-erb-B oncogene protein sequences. Nature 307:521–527
Fuhrer JP, Evans CH (1982) Rapid separation of biologically active syrian hamster lymphotoxin in high yield by size exclusion high performance liquid chromatography. J Chromatogr 248:427–433
Fullmer CS, Wasserman RH (1979) Analytical peptide mapping by high performance liquid chromatography. J Biol Chem 254:7208–7212
Geng X, Regnier FE (1984) Retention model for proteins in reversed phase liquid chromatography. J Chromatogr 296:15–30
Gerber GE, Anderegg RJ, Herlihy WC, Gray CP, Biemann K, Khorana HG (1979) Partial primary structure of bacteriorhodopsin: sequencing methods for membrane proteins. Proc Natl Acad Sci USA 76:227–231
Gooding KM, Schmuck MN (1984) Ion selectivity in the high performance cation-exchange chromatography of proteins. J Chromatogr 296:321–328
Gruber KA, Whitaker JM, Morris M (1979) Molecular weight separation of proteins and peptides with a new high pressure liquid chromatography column. Anal Biochem 97:176–183
Hearn MTW (1984) Reversed-phase high-performance liquid chromatography. In: Hirs CHW, Timasheff SN (eds) Methods in enzymology, vol 104. Academic Press, New York, pp 190–213
Hearn MTW, Regnier FE, Wehr CT (1982) HPLC of peptides and proteins. Am Lab (Fairfield Conn) 14:18–39
Hefti F (1982) High performance size exclusion chromatography: a buffer for the reliable determination of molecular weights of proteins. Anal Biochem 121:378–381
Heukeshoven J, Dernick R (1982) Reversed phase high performance liquid chromatography of virus proteins and other large hydrophobic proteins in formic acid containing solvents. J Chromatogr 252:241–254
Hewick RM, Hunkapiller MW, Hood LE, Dryer WJ (1981) A gas-liquid solid phase peptide and protein sequenator. J Biol Chem 256:7990–7997
Himmel ME, Squire PG (1981) High performance size exclusion chromatography of sea worm chlorocruorine and other large proteins, viruses and polysaccharides on a TSK G5000 SW preparative column. J Chromatogr 210:443–452
Hjerten S (1973) Some general aspects of hydrophobic interaction chromatography. J Chromatogr 87:325–331
Hunkapiller MW, Strickler JE, Wilson KJ (1984) Contemporary methodology for protein structure determination. Science 226:304–311
Imamura T, Konishi K, Yokoyama M, Konishi K (1979) High speed gel filtration of polypeptides in sodium dodecylsulfate. J Biochem 86:639–642
James S, Bennett HPJ (1985) Use of reversed-phase and ion-exchange batch extraction in the purification of the bovine pituitary peptides. J Chromatogr 326:329–338
Jenik RA, Porter JW (1981) High performance liquid chromatography of proteins by gel permeation chromatography. Anal Biochem 111:184–188
Johnsson A, Heldin C-H, Westermark B, Wasteson A (1982) Platelet-derived growth factor: identification of constituent polypeptide chains. Biochem Biophys Res Commun 104:66–74

Johnsson A, Heldin C-H, Wasteson A, Westermark B, Deuel TF, Huang JS, Seeburg PH, Gray A, Ullrich A, Scrace G, Stroobant P, Waterfield MD (1984) The c-sis gene encodes a precursor of the B chain of platelet-derived growth factor. EMBO J 3:921–928

Kamp RM, Bosserhoff A, Kamp D, Wittmann-Liebold (1984) Application of high performance liquid chromatographic techniques to the separation of ribosomal proteins of different organisms. J Chromatogr 317:181–192

Karsnas P, Moreno-Lopez J, Kristiansen T (1983) Bovine viral diarrhea virus: purification of surface proteins in detergent-containing buffers by fast protein liquid chromatography. J Chromatogr 266:643–649

Kato Y, Komiya K, Sasaki H, Hashimoto T (1980 a) Separation range and separation efficiency in high speed gelfiltrationon TSK-gel SW columns. J Chromatogr 190:297–303

Kato Y, Komiya K, Sasaki H, Hashimoto T (1980 b) High speed gelfiltration of proteins in sodium dodecyl sulphate aqueous solution on TSK-gel SW type. J Chromatogr 193:29–36

Kato Y, Komiya K, Sasaki H, Hashimoto T (1980 c) High speed gelfiltration of proteins in 6 M guanidine hydrochloride on TSK-gel SW columns. J Chromatogr 193:458–463

Kerlavage AR, Weitzmann CJ, Hasan T, Cooperman BS (1983) Reversed phase high performance liquid chromatography of *E. coli* ribosomal proteins. Characteristics of the separation of a complex protein mixture. J Chromatogr 266:225–237

Kopaciewicz W, Regnier FE (1982) Nonideal size exclusion chromatography of proteins: effects of pH at low ionic strength. Anal Biochem 126:8–16

Kopaciewicz W, Rounds MA, Fausnaugh J, Regnier FE (1983) Retention model for high performance ion-exchange chromatography. J Chromatogr 266:3–21

Kratzin H, Yang CY, Hilschmann N (1982) Chromatography and rechromatography in HPLC separation of peptides. In: Elzinga M (ed) Methods in protein sequence analysis. Humana Clifton, New Jersey, pp 427–434

Kratzin HD, Kruse T, Maywald F, Thinnes FP, Gotz H, Egert G, Pauly E, Friedrich J, Yang C-Y (1984) Primary structure of human class II histocompatibility antigens. Reversed phase high performance liquid chromatography for integral membrane proteins. J Chromatogr 297:1–11

Kumagaye KY, Takai M, Chino N, Sakakibari S (1985) Comparison of reversed-phase and cation-exchange high performance liquid chromatography for separating closely related peptides: separation of Asp^{76}-human parathyroid hormone (1-84) from Asn^{76}-human parathyroid hormone (1-84). J Chromatogr 327:327–332

Lambotte P, VanSnick J, Boon T (1984) Partial purification of a membrane glycoprotein antigen by high performance size exclusion chromatography without loss of antigenicity. J Chromatogr 297:139–145

Lewis RV, Fallon A, Stein S, Gibson KD, Udenfried S (1980) Supports for reversed phase high performance liquid chromatography of large proteins. Anal Biochem 104:153–159

Lindegren G, Lundstrom B, Kallman I, Hansson K-A (1984) Physical characteristics and properties of new chromatographic packing materials for the separation of peptides and proteins. J Chromatogr 296:83–95

Link GW, Keller PL, Banes AJ (1984) Methanol but not azide-H_2O as storage media favors subsequent binding to size exclusion packing material. 4th Int Symp on HPLC of proteins, peptides and polynucleotides. 10–12 December 1984, Baltimore (Abstract 110)

Ludi H, Hasselbach W (1984) State of aggregation of detergent-solubilized sarcoplasmic reticulum adenosine triphosphate investigated by high performance liquid chromatography. J Chromatogr 297:111–117

Mabuchi H, Nakahashi H (1981) Systematic separation of medium-sized biologically active peptides by high performance liquid chromatography. J Chromatogr 213:275–286

Mahoney WC (1982) Isolation of denatured proteins and peptides by high performance liquid chromatography. Effects of different perfluorinated acids, column length and large pore supports. Biochem Biophys Acta 704:284–289

Mahoney WC, Hermodson MA (1980) Separation of large denatured peptides by reversed phase high performance liquid chromatography. J Biol Chem 255:11199–11203
McGregor JL, Clezardin P, Manach M, Gronlund S, Dechavanne M (1985) Tandem separation of labelled human blood platelet glycoproteins by anion-exchange and gel fast protein liquid chromatography. J Chromatogr 326:179–190
Meek JL (1980) Prediction of peptide retention times in high pressure liquid chromatography on the basis of amino acid composition. Proc Natl Acad Sci USA 77:1632–1636
Meek JL, Rosetti ZL (1981) Factors affecting retention and resolution of peptides in high performance liquid chromatography. J Chromatogr 211:15–28
Muto N, Tan L (1985) Purification of oestrogen synthase by high-performance liquid chromatography. Two membrane bound enzymes from the human placenta. J Chromatogr 326:137–146
Nice EC, Lloyd CJ, Burgess AW (1984) The role of short microbore high performance liquid chromatography columns to protein separation and trace enrichment. J Chromatogr 296:153–170
O'Hare MJ, Capp MW, Nice EC, Cooke NHC, Archer BG (1982) Factors influencing chromatography of proteins on short alkylsilane-bonded large pore-size silicas. Anal Biochem 126:17–28
Pearson JD (1984) Micro HPLC-columns intended for submicrogram scale protein isolation. 4th Int Symp on HPLC of proteins, peptides and polynucleotides. 10–12 December 1984, Baltimore (Abstract 206)
Pearson JD, Mahoney WC, Hermodson MA, Regnier FE (1981) Reversed phase supports for the resolution of large denatured protein fragments. J Chromatogr 207:325–332
Pearson JD, Liu NT, Regnier FE (1982) The importance of silica type for reversed phase protein separations. Anal Biochem 124:217–230
Pfannkoch E, Lu KC, Regnier FE, Barth HH (1980) Characterization of some commercial high performance size-exclusion chromatography columns for water-soluble polymers. J Chromatogr Sci 180:430–441
Porath J, Sundberg L, Fornstedt N, Olsson I (1973) Salting-out in amphiphilic gels as a new approach to hydrophobic adsorption. Nature 245:465–466
Rabel FM (1980) Use and maintenance of microparticle high performance liquid chromatography columns. J Chromatogr Sci 18:394–408
Regnier FE (1984) High performance liquid chromatography of membrane proteins. In: Venter JC, Harrison LC (eds) Receptor purification procedures. Liss, New York, pp 61–75
Richter WO, Jacob B, Schwandt P (1983) Molecular weight determination of peptides by high performance gel permeation chromatography. Anal Biochem 133:288–291
Rivier JE (1978) Use of trialkyl ammonium phosphate (TAAP) buffers in reversed phase HPLC for high resolution and high recovery of peptides and proteins. J Liq Chromatogr 1:343–366
Rivier JE (1980) Evaluation of triethylammoniumphosphate and formate-acetonitrile mixtures as eluents for high performance gel permeation chromatography. J Chromatogr 202:211–222
Rivier JE, McClintock R (1983) Reversed phase high performance liquid chromatography of insulins from different species. J Chromatogr 268:112–119
Rivier JE, Rivier C, Spiess J, Vale W (1982) High performance liquid chromatographic purification of peptide hormones: ovine hypothalamic amumine. Anal Biochem 127:258–266
Rivier JE, McClintock R, Galyean R, Anderson (1984) Reversed phase high performance liquid chromatography: preparative purification of synthetic peptides. J Chromatogr 288:303–328
Sasagawa T, Okuyama T, Teller DC (1982) Prediction of peptide retention times in reversed phase high performance liquid chromatography during linear gradient elution. J Chromatogr 240:329–340
Spindel E, Wurtman RJ (1979) Reversed phase ion-pair separation of thyrotropin-releasing hormone and some analogs. J Chromatogr 175:198–201

Su S-W, Grego B, Niven B, Hearn MTW (1981) Analysis of group retention contributions for peptides separated by reversed phase high performance liquid chromatography. J Liq Chromatogr 4:1745–1764

Swergold GD, Rubin CS (1983) High performance gel permeation chromatography of polypeptides in a volatile solvent: rapid resolution and molecular weight estimations of proteins and peptides on a column of TSK-3000-PW. Anal Biochem 131:295–300

Takagi T, Takeda K, Okuno T (1981) Effect of salt concentration on the elution properties of complexes formed between sodium dodecyl sulphate and protein polypeptides in high performance silica gel chromatography. J Chromatogr 206:201–208

Takahashi N, Ishioka N, Takahashi Y, Putman FW (1985) Automated tandem high-performance liquid chromatographic system for separation of extremely complex peptide mixtures. J Chromatogr 326:407–418

Tandy NE, Dilley RA, Regnier FE (1983) High performance liquid chromatographic purification of the hydrophobic ω subunit of the chloroplast energy coupling complex. J Chromatogr 266:599–608

Tarr GE, Crabb JW (1983) Reversed phase high performance liquid chromatography of hydrophobic proteins and fragments thereof. Anal Biochem 131:99–107

Tempst P, Hunkapiller MW, Hood LE (1984) Separation of peptides by reversed phase high performance liquid chromatography using propyl- and cyanopropylsilyl supports. Anal Biochem 137:188–195

Ui N (1979) Rapid estimation of the molecular weights of protein polypeptide chains using high pressure liquid chromatography in 6 M guanidine hydrochloride. Anal Biochem 97:65–71

Van der Zee R, Welling GW (1982) Molecular sieving during reversed phase high performance liquid chromatography of proteins. J Chromatogr 244:134–136

Vanecek G, Regnier FE (1980) Variables in the high performance anion-exchange chromatography of proteins. Anal Biochem 109:345–353

Waterfield MD, Scrace GT, Whittle N, Stroobant P, Johnsson A, Wasteson A, Westermark B, Heldin C-H, Huang JS, Deuel TF (1983) Platelet-derived growth factor is structurally related to the putative transforming protein p 28sis of simian sarcoma virus. Nature 304:35–39

Wehr CT, Correia L, Abbott SR (1982a) Evaluation of stationary and mobile phases for reversed phase high performance liquid chromatography of peptides. J Chromatogr Sci 20:114–119

Wehr CT, Cunico RL, Ott GS, Shore VG (1982b) Preparative size exclusion chromatography of human serum apolipoproteins using an analytical liquid chromatograph. Anal Biochem 125:386–394

Welling GW, Nijmeijer JRJ, Van Der Zee R, Groen G, Wilterdink JB, Welling-Wester S (1984) Isolation of detergent extracted sendai virus proteins by gel filtration, ion-exchange and reversed phase high performance liquid chromatography and the effect on immunological activity. J Chromatogr 297:101–109

Welling GW, Groen G, Slopsema K, Welling-Wester S (1985) Combined size-exclusion and reversed-phase high-performance liquid chromatography of a detergent extract of Sendai virus. J Chromatogr 326:173–178

Wilson KJ, Honnegger A, Stotzel RP, Hughes GJ (1981a) The behavior of peptides on reversed phase supports during high pressure liquid chromatography. Biochem J 199:31–41

Wilson KJ, Honnegger A, Hughes GJ (1981b) Comparison of buffers and detection systems for high pressure liquid chromatography of peptide mixtures. Biochem J 199:43–51

Wilson KJ, Van Wieringen E, Klauser S, Berchtold MW (1982) Comparison of the high performance liquid chromatography of peptides and proteins on 100 and 300 Å reversed phase supports. J Chromatogr 237:407–416

Monitoring Polyamines in Plant Tissues by High Performance Liquid Chromatography

M. A. Smith and P. J. Davies

1 Introduction

The commonly studied amines are the diamines, diaminopropane, cadaverine and putrescine, the triamine, spermidine and the tetraamine, spermine. Although the physiological role of polyamines remains unclear, they are considered important in the growth of plants, animals, and microorganisms (Bachrach 1973). The function and biochemistry of polyamines, and other naturally occurring amines in plants are discussed in several recent reviews (Smith 1980; Galston 1983; Slocum et al. 1984; Smith 1985). In plants they have been associated with many cell processes, including cell division and differentiation, membrane stability, pH and osmotic stress, retardation of senescence, and the synthesis of nucleic acids and proteins. In addition, aliphatic polyamines, as well as aromatic amines, have been found as constituents of alkaloids and conjugates of cinnamic acids (see reviews by Guggisberg and Hesse 1983; Smith et al. 1983; Slocum et al. 1984; Smith 1985).

Broad interest in following the changes in the levels of a wide spectrum of amines in plants during growth and development has required the development of sensitive and routine methods for their analysis. Several techniques have been employed to analyze plant tissues for polyamines, including gas chromatography (GC) (Anderson and Martin 1973), paper electrophoresis (PE) (Fracassini et al. 1980; Bagni 1970), thin-layer electrophoresis (TLE) (Anderson and Martin 1973), thin-layer chromatography (TLC) (Smith and Wilshire 1975; Smith and Best 1977; Flores and Galston 1982), high performance liquid chromatography (HPLC) by ion exchange (Adlakha and Villanueva 1980; Villanueva et al. 1985, 1986a, b) and more recently, reverse phase HPLC of polyamine derivatives (Flores and Galston 1982; Smith and Davies 1985) and ion-pair reverse phase HPLC of underivatized polyamines (Greenberg and Cohen 1985). Reverse phase HPLC of derivatized polyamines has also been followed by solid probe mass spectrometry (Roberts et al. 1985).

Four sensitive methods have been used to analyze plant tissue for polyamines, including reverse phase HPLC of dansyl- or benzoyl derivatives, with fluorescence and UV monitoring respectively, and ion-pairing reverse phase HPLC or ion-exchange chromatography, with fluorogenic detection following post-column derivatization with o-phthalaldehyde (OPA). HPLC has been extended to the analysis of a large number of other naturally occurring amines in plant tissues, including polyamine homologs, heterocyclic and aromatic amines. HPLC also offers the advantages over TLC, TLE, or PE of superior resolution and easier quantification of polyamines.

2 High Performance Liquid Chromatography of Dansylamines

2.1 Dansyl Derivatives

Dansyl chloride (5-dimethylaminonaphthalene-1-sulfonyl chloride) has been the most widely used reagent for derivatization of polyamines (Seiler 1983 a) prior to separation by PE, TLC, or HPLC. The methods for dansylation of amines in plant tissue extracts have been modified from the methods of Seiler and Wiechmann (1967, 1970), Seiler (1971), and Seiler et al. (1978).

Dansyl chloride reacts stoichiometrically with primary and secondary amines (Fig. 1). If these substances have more than one amino group, each is derivatized by the reagent (Seiler and Wiechmann 1970). Aside from the commonly studied polyamines (putrescine, spermidine, cadaverine, and spermine), dansyl chloride reacts with many extractable compounds from plants, including other aliphatic amines, aromatic and heterocyclic amines, alkaloids, ammonia, amino acids, purine bases, phenols, aliphatic alcohols, and some sugars (Seiler and Wiechmann 1970). Despite the nonspecificity of dansyl chloride, detection of polyamines in the presence of competing compounds can be accomplished easily, since the concentrations of polyamines in plant tissues are usually greater than many other reacting substances. Interference from dansyl amino acids is avoided, since they remain in the aqueous phase when dansyl polyamines are extracted from the reaction mixture with an organic solvent.

Dansylation is performed in the dark since the derivatives are light-sensitive, but dim light in the laboratory is tolerable (Seiler 1983 a). In subdued laboratory light, dansyl derivatives show a 50% decrease in fluorescence after 1 week, but direct sunlight causes complete loss of fluorescence after 60 min (Smith and Best 1977). When dansylamines are stored in methanol at 4 °C in the dark, they are stable for up to a month without substantial decomposition ($<10\%$) (Kabra et al. 1986).

A general protocol for the extraction, dansylation and detection of polyamines is depicted in Fig. 2.

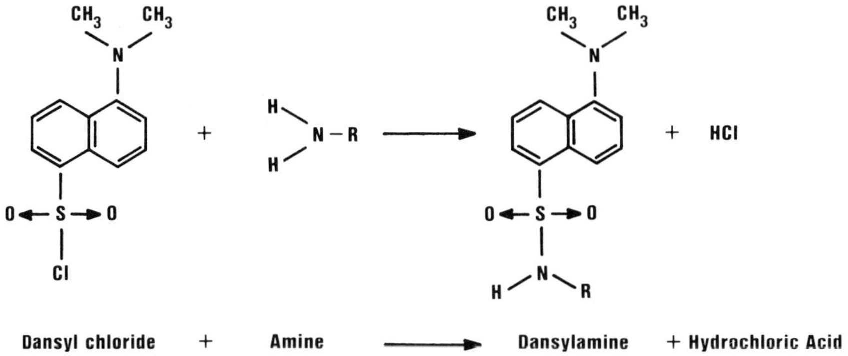

Fig. 1. Reaction of dansyl chloride with a primary amine

Extract in weak acid solution (e.g., 0.2 N PCA) on ice (100 mg tissue ml^{-1})

Let stand in the cold 30 to 60 min

Centrifuge

Save supernatant

To 100 µl add 200 µl saturated Na_2CO_3 + 400 µl dansyl–Cl (5 to 30 mg ml^{-1}) in acetone

Incubate in dark at 60 °C for 1 h or at room temperature for 16 h overnight or sonicate 2 to 3 h

Add 100 µl proline (100 mg/ml)

Incubate 10 to 30 min, or sonicate 3 min

Extract with 500 µl or more of toluene or ethyl acetate by vortexing 15–30 s

Centrifuge to separate phases

Discard aqueous phase and dry organic phase under N_2

Re-dissolve in methanol or acetonitrile

Pass sample through 0.22 µm pore size filter

Dilute if necessary and apply 20 to 50 µl to a C_{18} column of the HPLC

Elute with appropriate gradient (e.g., 60 to 95% water/methanol over 23 min)

Detect with fluorescence detector equipped with a flow through cell using optimal excitation and emission wavelengths

Use an online integrator to record peaks and to integrate peak areas

Fig. 2. Scheme for the extraction, dansylation, and detection of polyamines

2.2 Reagents and Stock Solutions

Polyamines are obtained as hydrochloride salts. Standard solutions of putrescine, cadaverine, hexanediamine, spermidine, or spermine are made up in 1 to 10 mM concentrations in water, or in low concentrations of acids, e.g., 0.5% perchloric acid (PCA) or 0.1 N hydrochloric acid (HCl). If stored in plastic tubes at −20 °C they are stable for at least 2 months or more (Goren et al. 1982; Flores and Galston 1982; Kabra et al. 1986; Bontemps et al 1984b).

Plastic containers are used for storage because polyamines adsorb to the surface of glass. They also bind to certain membranes and cellophane dialysis tubing (Grettie et al. 1978). Therefore membrane filters should be tested for quantitative analysis.

Impure dansyl chloride produces poor results. The reagent can be purified on a column of silica gel 60 as described by Seiler (1983a). Solutions of dansyl chloride in acetone (10 mg ml^{-1}) can be stored for 24 h at 4 °C in the dark (Kabra et al. 1986).

2.3 Extraction of Polyamines

Polyamines are extracted from tissue in low concentrations of cold trichloroacetic acid (TCA), HCl/TCA mixture, or PCA, and then incubated in the medium for 30 to 60 min. Depending on the tissue, the homogenate is centrifuged at 3000 to 26,000 *g* in the cold for 10 to 30 min, and the supernatant is saved for polyamine analysis. If TCA is used for homogenation, it is removed by ether extraction prior to dansylation, since it interferes with derivatization (Smith and Best 1977). PCA is most often the extraction medium of choice.

Fresh tissue is homogenized in cold 0.2 N PCA (100 mg tissue ml^{-1} acid) in a mortar with a pestle, or for small samples, in 5-ml centrifuge tubes with ground glass conical bases with matching pestles. To prevent polyamines from binding to the walls of glass apparatus used in extraction and subsequent procedure, all glassware is siliconized, according to the directions provided with the siliconizing fluid (Aquasil, Rockford, IL, USA). Organic contaminants on glassware may be removed by a heat pretreatment at approximately 500 °C, on the cleaning cycle in a domestic oven. Hexanediamine (1,6 diaminohexane, an unnatural polyamine) is added as an internal standard, before the tissue is ground, to a concentration similar to the endogenous polyamine levels estimated to be present in the tissue. The samples are incubated on ice for 30 to 60 min, and then centrifuged at 4 °C for 20 min. The supernatant is saved for polyamine analysis (Smith and Davies 1985).

2.4 Purification of Extracts

Crude extracts from plant tissues can be dansylated without pre-purification. Should purification be necessary, the following purification procedures, adapted from Grettie et al. (1978), can be applied to extracts treated by any pre- or post-column derivatization method. The acid extract is adjusted to pH 9.0 with 1 N NaOH, and then applied to a small column of 0.5 g of silica gel powder (60 to 200 mesh, J. T. Baker, Phillipsburg, NJ, USA). The column is washed with distilled water (adjusted to pH 8.0 with diluted NaOH) and then eluted with 30 ml of 0.03 N HCl. The eluate is dried at 45 °C and redissolved in 1.0 ml of 0.03 N HCl (Cohen et al. 1981). Similarly, the purification can be done on silica Sep-Paks

(Waters Associate, Milford, MA, USA), as described by Seiler (1983 a) and Brossat et al. (1983). Two ml of supernatant, adjusted to pH 9.0, is loaded on to a Sep-Pak cartridge with a syringe. After washing with 5 ml of water, polyamines are eluted with 10 ml of 0.1 N HCl. Silica gel columns can also be used to purify commercial radiolabeled polyamines (Grettie et al. 1978).

2.5 Dansylation of Amines in a Sample

Generally, an aliquot of the supernatant of a crude plant extract (100 to 200 μl, representing about 10 to 50 mg of tissue), is added to 400 μl of dansyl chloride in acetone (5 to 30 mg ml^{-1}) and 200 μl of saturated Na_2CO_3. The saturated Na_2CO_3 is used to produce an alkaline medium for labeling, which is optimal at pH of 9.5 to 10.5 (Seiler and Wiechmann 1970). The reaction mixture is added to amber or aluminum foil-covered 3- to 5-ml Reacti-vials (Pierce, Rockford, IL, USA) and incubated at 60 °C for 1 hour in a heating block (Smith and Davies 1985). However, phase separation may occur under these conditions (Seiler 1983 a), so small magnetic spin bars should be placed in each vial and incubation is done in a heating block equipped with a magnetic stirrer (Pierce, Rockford, IL, USA). Alternatively, the reaction is allowed to proceed in the dark for 16 h or overnight at room temperature. Short incubation procedures, which have not been reported for plant analysis, are also possible. Sonication of the reaction mixture at room temperature for 2 to 3 h (Heby and Andersson 1978; Seiler 1983 a) or incubation at 70 °C for 10 min (Kabra et al. 1986) have been reported.

To remove the excess dansyl chloride, 100 μl of proline (100 mg ml^{-1}) is added to the mixture and incubated for 30 min at room temperature. An incubation time as short as 10 min has been used also (Kabra et al. 1986). Heby and Andersson (1978) reported the reaction to be quantitative after 3 min of sonication.

Dansylated polyamines are extracted from the alkaline medium with 500 μl or more of toluene or ethyl acetate by vortexing for 15–30 s. After brief centrifugation, the aqueous phase which contains dansylated amino acids is discarded. The organic phase containing the derivatized polyamines is dried under a stream of N_2. The residue is then redissolved in a solvent compatible with the HPLC column. Commonly used solvents are acetonitrile or methanol.

An alternative procedure to remove polar reaction products uses a Bond-Elut C_{18} column (Analytichem International, Meltrose Park, IL, USA) (Kabra et al. 1986). The dansyl chloride/amine reaction mixture is cooled to room temperature, and then transferred to a column on a vacuum apparatus. After the column drains, it is washed with two column volumes of water before polyamines are eluted with 500 μl of methanol. The sample is then ready to be loaded on to the HPLC.

A standard mixture of polyamines, and blank samples (containing the reaction mixture and extraction medium with no polyamines) are prepared as described above.

2.6 Chromatographic Separation of Dansylamines

Separations can be achieved on any high resolution (5- to 10 μm sized spherical particles) reverse phase C_{18} (15–30 cm × 2.2 to 5 mm i.d.) (Saeki et al. 1978; Seiler 1983a; Hayman et al. 1985; Smith and Davies 1985; Kabra et al. 1986). In this and similar HPLC procedures, a guard column packed with a similar C_{18} material should be used to protect the analytical column. For additional protection, in all methods, samples should be ultrafiltered through 0.22-μm pore size nylon membranes before application to the HPLC. For small sample volumes, centrifugal ultrafiltration tubes (Rainin, Woburn, MA, USA) are most convenient. The samples are diluted and loaded in low amine concentrations, nanomole to picomole levels in 20 to 50 μl of solvent, depending on the sensitivity of the fluorescence detector used. A fixed-volume 20 to 50-μl injector loop, filled to capacity, is recommended for sample application.

Polyamines are eluted from 5-μm reverse phase C_{18} columns (25 cm × 4.6 or 5 mm i.d.) using a simple linear gradient of 60% methanol in water to 95% methanol over 23 min at a flow rate of 1 ml min^{-1} (Smith and Davies 1985). We found best separations without buffering, as acid pH in our solvent system caused the formation of double peaks. In all HPLC methods, solvents should be HPLC grade (from, e.g., J. T. Baker, Phillipsburg, NJ, USA or Burdick and Jackson, Muskegon, MI, USA), and prior to use on HPLC they should be filtered through 0.45-μm pore size Millipore membranes and degassed in vacuo. Separations are also possible with appropriate gradients of acetonitrile in water (Saeki et al. 1978) or acetonitrile in 10 mM phosphate buffer, pH 4.4 (Kabra et al. 1986).

Dansylamines are monitored with a fluorescence detector, equipped with a flow-through cell. To select optimal excitation and emission wavelengths for monitoring, the solvent flow is stopped, and each spectrum is scanned while the flow cell contains a dansylated sample. For various instruments, the excitation wavelengths to activate fluoresence of the dansyl derivatives have been reported to be as low as 333 nm and as high as 365 nm. To measure fluorescence, optimal emission wavelengths have been chosen from 485 to 522 nm. We found optimal wavelengths at 365 nm and 510 nm for excitation and emission, respectively, using a fluorescence spectrophotometer (model 650-10LC, Perkin Elmer) with an 8-μl cell.

A typical chromatogram of polyamine standards and derivatives of plant extracts are shown in Fig. 3A and B, respectively. With a gradient elution of 60% methanol in water to 95% methanol over 23 min, excellent resolution of putrescine, cadaverine, hexanediamine (internal standard), spermidine and spermine is achieved within 27 min ($V_0 = 2.03$ ml). Faster analysis of dansyl derivatives can be achieved by changing the organic solvent in the mobile phase, steepening the elution gradient and/or by using a higher resolution reverse phase column (3-μm particle size) (Saeki et al. 1978; Bontemps et al. 1984a, b; Kabra et al. 1986). Some investigators have included ion-pairing agents such as octanesulfonate or heptanesulfonate in the mobile phase to effect solute separation (Brown et al. 1979; Brown et al. 1982, Brossat et al. 1983; Bontemps et al. 1984a, b). For analysis of the dansyl derivatives of the commonly studied aliphatic amines (putrescine, cadaverine, spermidine, and spermine), this mecha-

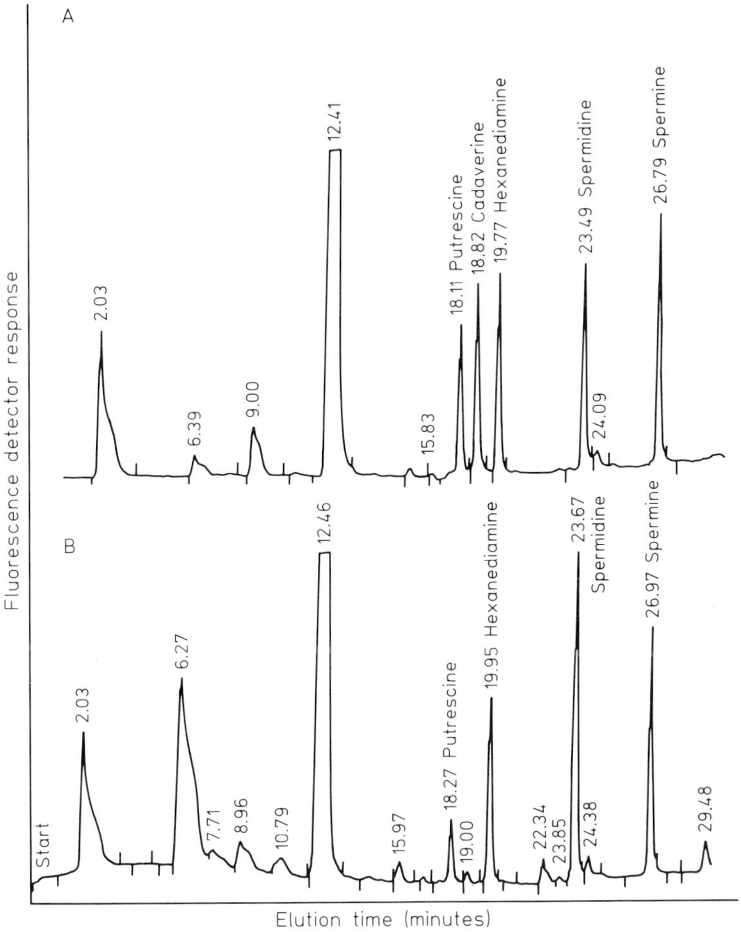

Fig. 3 A, B. Separation of dansylamines (**A**)standards (2.5 picomoles) and (**B**) from extracts of pea buds (representing 5 µg of tissue), by reverse phase HPLC on a 5-µm C_{18} column, using a gradient of water/methanol (60 to 95% over 23 min) at a flow rate of 1 ml min^{-1}. Sample volume was 20 µl. Fluorescence detection with 365 nm for excitation and 510 nm for emission. (After Smith and Davies 1985)

nism has not been necessary for extracts from plants, as excellent resolution is obtained without ion pairing.

Several side products elute well before the polyamine derivatives (Fig. 3 A and B). These products are always produced, even in blank preparations, in the absence of sample, under the conditions for dansylation (Seiler and Wiechmann 1970). Some of these compounds have been identified by Seiler et al. (1978), using HPLC, to be dansyl methylamine, dansyl dimethylamine, dansyl ammonia, and dansyl hydroxide. The amount of these substances depends on the amount of dansyl chloride used, pH, and temperature of the reaction mixture, the length of the reaction time (Seiler 1971), and pre-chromatographic purification.

After elution of spermine, the organic component of the mobile phase is increased to 100% for a few minutes to remove strongly adsorbed contaminants. Then the gradient is gradually lowered to the initial solvent conditions, and the column is re-equilibrated for at least 5 min in preparation for a new sample injection. Analysis of the precursor to putrescine, agmatine, in plant extracts, was not successfully achieved, due probably to the instability of the derivative. Agmatine can produce side reactions with dansyl chloride as a result of fragmentation (Seiler and Wiechmann 1970; Hayman et al. 1985). From a reaction mixture of agmatine with 4 M dansyl chloride, three fragmentation products were recovered, mono-dansyl agmatine, bis-dansyl agmatine, and bis-dansyl putrescine (Seiler and Wiechmann 1970).

Separation of a wider variety of naturally occurring amines, aside from the commonly studied aliphatic compounds, is possible, with the appropriate mobile phase and elution gradient, through HPLC of their dansyl dervatives. Polyamine homologs, nor-spermidine and nor-spermine have been detected in extracts of *Euglena gracilis* by this method (Kneifel et al. 1978). Derivatives of standards of heterocyclic and aromatic amines which are endogenous to plants have been resolved by HPLC (Seiler et al. 1978; Hayman et al. 1985), but application of the procedures to plant extracts have not been published. Hayman et al. (1985) using a series of ten isocratic elution steps, with increasing concentrations of acetonitrile in 17.5 mM phosphate buffer (pH 7.17), at 35 °C, and a flow rate of 1.5 ml min^{-1}, separated derivatives of 29 out of 45 standard amines, simultaneously, in 78 min on a 5-μm particle size, 25 cm × 5 mm i.d. (C_{18}) column. They report that their results for plant tissue are encouraging.

To identify unknown amines in plant extracts, derivatized standards of endogenous compounds are chromatographed individually, to determine characteristic retention times. Then, samples are analyzed and component peaks identified according to similarities to retention times of the standards. Additional confirmation of identity is obtained by derivatizing and chromatographing a sample with added standards, and checking for increases in peak areas. Alternatively, for more definitive identification, the amine fractions can be collected and analyzed by mass spectrometry (Seiler et al. 1978; Creveling and Daly 1971).

Recoveries are determined by adding ^{14}C-labeled polyamines to the homogenates and measuring radioactivity after chromatographic separation. Another approach is to add known amounts of unlabeled standards of the endogenous amines to one of two similar samples, and then calculate the recovery of the standards based on the difference in the amounts of amines detected in both samples after chromatography. Using the latter method we determined recoveries of 90% or better for the internal standard, 85 to 99% for putrescine, 81 to 96% for spermidine and 76 to 94% for spermine (Smith and Davies, unpublished).

2.7 Quantification

Quantification may be easily achieved by internal standardization, since unnatural amines are readily available, which may be used as reference compunds. Hexanediamine (1,6-diaminohexane) is often used as an internal standard because it is an unnatural amine, it resolves well from derivatives of endogenous amines, it elutes near amines of interest, and it is stable under storage conditions described earlier. Another suitable standard is 1,7 diaminoheptane.

Sample polyamines can be quantified from calibration curves generated for each endogenous amine in the tissue. A known amount of internal standard and increasing concentrations of standards of the endogenous polyamines, within the range of the amounts present in the tissue, are derivatized and chromatographed under the conditions for sample analysis. Area ratios of an endogenous amine to the internal standard are determined and plotted against corresponding ratios of concentrations (Smith and Davies 1985). The equations for the curves are determined by linear regression.

Alternatively, endogenous amine concentrations may be calculated relative to an internal standard by a simple equation and procedure. First, however, a fluorescence response factor is determined for each endogenous amine. Known amounts of standards of endogenous amines are mixed with a known quantity of an internal standard and chromatographed under assay conditions. Areas of the peaks for each compound are integrated and then the fluorescence response factor (FRF) for each endogenous amine is determined as follows:

$$\mathrm{FRF} = \frac{\text{Peak area of amine}}{\text{Peak area of internal standard}} \times \frac{\text{Concentration of internal standard}}{\text{Concentration of amine}}.$$

The unknown amount of an endogenous amine in a sample containing a known concentration of internal standard is determined by the equation below, which includes the FRF to correct for variations in assay conditions

$$\text{Amount of amine} = \frac{\text{Peak area of amine}}{\text{Peak area of internal standard}} \times \frac{1}{\mathrm{FRF}} \times \text{Added amount of internal standard}.$$

Verification of this equation should be done by plotting area ratios against corresponding concentration ratios to show the intercept passes through zero. Otherwise the equation includes a value other than zero for the intercept (see Smith and Davies 1985). The slope of the curve is the FRF.

Fluorescence monitoring of polyamines through their dansyl derivatives is a rapid and highly sensitive method with limits of detection at the femtomolar level (Brown et al. 1982; Bontemps et al. 1984b; Smith and Davies 1985; Kabra et al. 1986).

3 High Performance Liquid Chromatography of Benzoylamines

For HPLC analysis of polyamines with detection by UV absorption, pre-column derivatization with benzoyl chloride has been adapted from Redmond and Tseng (1979). The principles for benzoylation (Fig. 4) are the same as for dansylation,

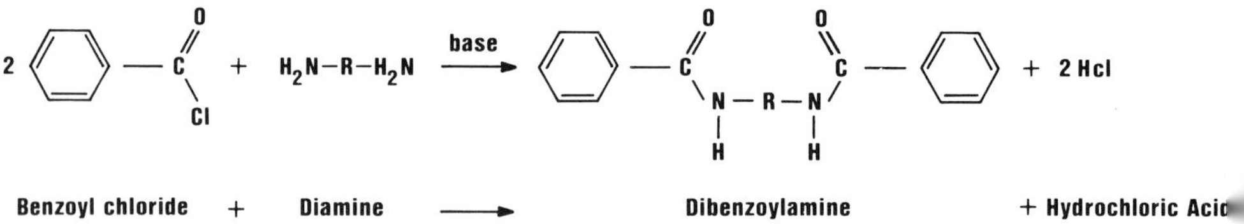

Fig. 4. Reaction of benzoyl chloride with a diamine

since acid chlorides react similarly (Seiler 1983a). However, the benzoylamines are not sensitive to light. With the exceptions of spermidine and agmatine, derivatives in methanol are stable for several months at −20 °C (Flores and Galston 1982).

3.1 Benzoylation of Amines in a Sample

Acid extracts are prepared as described earlier. Aliquots of 0.1 to 1 ml of the supernatant are added to 1 ml of 2 N NaOH with 5 to 10 µl of benzoyl chloride. The mixture is incubated at room temperature for 20 min, and then the reaction is terminated by the addition of 2 ml of saturated NaCl. Benzoylamines are then extracted with 2 ml of diethyl ether. After centrifugation, the ether layer is removed and dried under a stream of nitrogen. The residue is redissolved in 100 to 200 µl of methanol or acetonitrile (Redmond and Tseng 1979; Flores and Galston 1982).

3.2 Chromatographic Separation of Benzoylamines

The benzoyl derivatives are separated on C_{18} reverse phase columns, using either isocratic or gradient elutions with acetonitrile or methanol in the mobile phase (Flores and Galston 1982; Roberts et al. 1985). The eluates from the column are monitored by a UV detector with an absorbance wavelength of 254 nm.

Benzoylated standards of putrescine, cadaverine, agmatine, nor-spermidine, spermidine, homo-spermidine, and spermine are separated on a 5-µm, 25 cm × 4.6 mm i.d. (C_{18}) column in less than 20 min by isocratic elution with 64% methanol at a flow rate of 1 ml min^{-1} (Flores and Galston 1982). Derivatives of plant extracts are resolvable by isocratic elution, using organic solvent strengths between 60 and 65% (Fig. 5). The identities of the endogenous amine peaks in Fig. 5 were determined by retention times and co-chromatography with amine standards. Although a peak for agmatine is shown, detection of the derivative is not very reproducible (Flores, personal communication). In addition to common amines, HPLC of benzoylated derivatives has been used to detect the polyamine homolog, sym-nor-spermidine, in *Nicotiana tabacum* (Madsen et al. 1985).

Solid probe mass spectrometry has also been used to confirm the identities of putrescine, spermidine, and spermine from plant tissue after their separation as

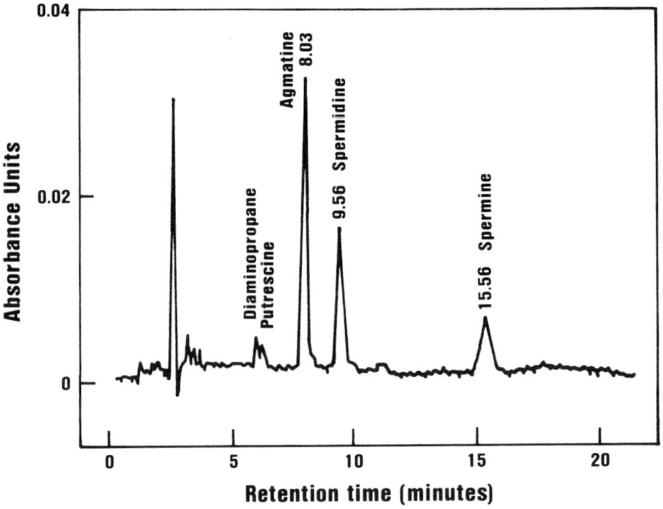

Fig. 5. Separation of benzoyl derivatives of polyamines from extracts of amaranth cotyledons (representing 10 mg of tissue) by reverse phase HPLC on a 5-μm C_{18} column, using isocratic elution with 64% methanol, followed by UV detection at 254 nm. Sample volume was 20 μl. (After Flores and Galston 1982)

benzoyl derivatives on a reverse phase C_{18} column (Roberts et al. 1984, 1985). Polyamines were resolved with an elution gradient of methanol : water : acetonitrile (8 : 11 : 1) changed to a ratio of 14 : 5 : 1 over 20 min. Recoveries of ^{14}C-labeled amines added to tissue extracts were also determined. For putrescine, spermidine, and spermine, recoveries were 94 ± 2, 83 ± 2, and 70 ± 2%, respectively. For unlabeled hexanediamine, the recovery was 83 ± 2.9%.

The preparation of the benzoylamines is a simple and rapid procedure. It is also inexpensive, since only 5 to 10 μl of benzoyl chloride is needed to derivatize a sample. HPLC of benzoylamines for polyamine analysis is a convenient method due to the general availability of UV detectors. Sensitivity, though not as good as for dansyl derivatives, is well below that which is needed for the analysis of plant tissues for polyamines. Detection limits between 0.01 and 1 nmol have been reported (Flores and Galston 1982).

4 Ion-Pair Reverse Phase Chromatography of Underivatized Polyamines

A third HPLC approach for polyamine analysis in biological tissues consist of ion-pair reverse phase liquid chromatography of underivatized polyamines with post-column derivatization with o-phthalaldehyde (OPA) (Wagner et al. 1982; Seiler 1983 b; Seiler and Knödgen 1985; Greenberg and Cohen 1985). Octanesulfonate or heptanesulfonate is used as an ion-pairing agent to neutralize the charges on the polyamines. As the polar polyamines become nonpolar their retention on the reverse phase column is increased. The retention and selectivity of solutes are controllable by the concentration and hydrophobicity of the ion-pairing agent, the nature of the organic solvent of the mobile phase, pH, and temperature (Cooke and Olsen 1979; Wagner et al. 1982). Thus, several approaches may be explored to improve separations. In addition, a large number of compounds in a single sample, including neutral and charged solutes, can be resolved (Cooke and Olsen 1979).

4.1 o-Phthalaldehyde Derivatives

OPA with 2-mercaptoethanol or ethanethiol reacts rapidly with primary amines to form strong fluorescent derivatives, 1-alkylthio-2-alkyl-substituted isoindoles, as shown in Fig. 6 (Simons and Johnson 1976). Thus, the reagent can be used for pre- or post-column derivatization of polyamines. Since OPA derivatives are not very stable, post-column derivatization is preferred. Derivatives formed with ethanethiol, however, are more stable than those produced with 2-mercaptoethanol (Skaaden and Greibrokk 1982).

4.2 Sample Preparation

Tissues are extracted as described earlier, diluted if necessary, and then applied to the column of the HPLC (Seiler 1983 b; Wagner et al. 1982). In ion-pairing reverse phase HPLC of crude extracts a large number of peaks of unknown substances appear on the chromatogram that may interfere with the determination of compounds of interest (Seiler 1983 b). Thus a clean-up step could be required. Crude PCA extracts of plant tissue yield many compounds with primary amine groups which prevent determinations of 1-aminocyclopropane-1-carboxylate (ACC) and methionine simultaneously with S-adenosylmethionine (SAM) and decarboxylated S-adenosylmethionine (dSAM) and polyamines (Greenberg and Cohen 1985). Therefore, extracts are fractionated on 1 × 5 cm columns of SP-Sephadex Pharmacia Fine Chemicals, Picataway NJ, USA prior to HPLC. PCA extracts are adjusted to pH 3 to 4 with KOH, chilled on ice for 30 min and then cen-

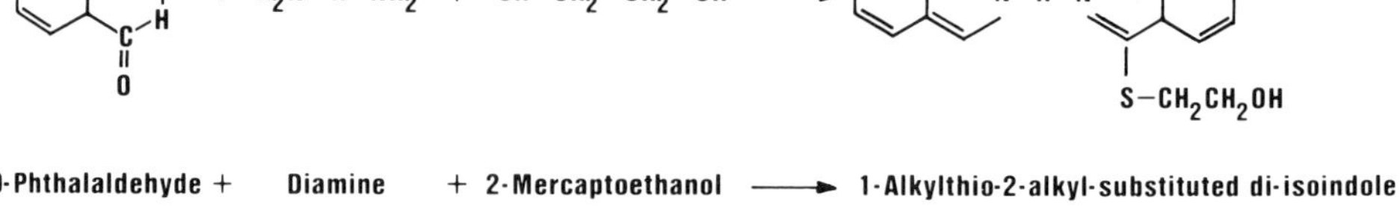

Fig. 6. Reaction of o-phthalaldehyde (OPA) and 2-mercaptoethanol with a diamine. Two molecules each of OPA and 2-mercaptoethanol are needed to form the di-isoindole derivative

trifuged. The supernatant containing $KClO_4$ is discarded and the precipitate is treated with cold 10 mM HCl and centrifuged. Then the latter supernatant is applied to the Sephadex column. Methionine and ACC are eluted with 35 ml of 50 mM HCl, putrescine with 35 ml of 150 mM HCl and SAM, dSAM and polyamines with 40 ml of 0.5 M HCl. Fractions are dried in vacuo at 32 °C and redissolved in an appropriate volume of 10 mM HCl. Then the methionine fraction and the polyamine/SAM fraction are analyzed separately by HPLC under different chromatographic conditions. Unfractionated PCA extracts of plant tissues can be applied directly to the HPLC for routine determinations of putrescine, spermidine, spermine, and SAM, although frequent changing of the guard column and cleaning of the analytical column are required to remove lipophilic substances.

4.3 Chromatographic Separation of Underivatized Polyamines

For ion-pair reverse phase HPLC of polyamines, extracts are applied to C_{18} columns with 4-μm particle size (15 cm × 3.9 mm i.d.) or 10-μm particle size (30 cm × 3.9 mm i.d.) and chromatographed at room temperature. Solutes are resolved by gradient elution, using the following mobile phases: (a) 0.1 M NaH_2PO_4, 8 mM heptanesulfonic acid, 0.1 mM Na_2EDTA (pH 2.5) and acetonitrile (132:1, v/v) and (b) 0.2 M NaH_2PO_4, 8 mM heptanesulfonic acid (pH 3.1) and acetonitrile (7:3, v/v). The gradient is started with 20% (b) and increased to 60% (b) in 21 min, using a flow rate of 1 or 1.5 ml min^{-1}, depending on the column used (Greenberg and Cohen 1985).

4.4 Post-Column Derivatization with OPA

For post-column derivatization, the OPA reagent is prepared as follows: A liter solution of 0.5 M boric acid is adjusted to pH 10.4 with 45 g of KOH. To this solution, 2.5 ml of 2-mercaptoethanol, 800 mg of OPA in 10 ml of ethanol and 3 ml of Brij-35 (30% solution in water) are added (Wagner et al. 1982; Benson and Hare 1975). The reagent can be stored for 2 days in dark bottles (Seiler 1983b). The mixture is pumped at a flow rate equal to that of the mobile phase into the column effluent line through a T-tube where it reacts with amines. Then the derivatives are transferred to a flow cell of an attached fluorometer (Wagner et al. 1982).

4.5 Detection of OPA Derivatives

Excitation wavelengths at 334–346 nm and emission at 425 nm are used for fluorogenic monitoring of the isoindole derivatives (Greenberg and Cohen 1985). By adding the OPA reagent mixture to the effluent line after sending the eluent through a UV monitor, S-adenosyl methionine (SAM) and decarboxylated SAM are detected by UV absorbance at 254 nm with subsequent detection of polyamines as isoindole derivatives by fluorometry. The polyamines do not absorb

UV and the SAM analogs react very weakly, if at all, with OPA (Wagner et al. 1982). However, methionine reacts with the reagent and is therefore detected by derivatization after HPLC. Reportedly, methionine and ACC can be separated and detected by this method, but the results for plant extracts were not shown (Greenberg and Cohen 1985).

A feature of the ion-pairing approach is the potential it offers for rapid and simultaneous determinations of polyamines and other important metabolites in a single sample. Separations depend on the version of the method. The procedure of Greenberg and Cohen (1985), for example, allows the resolution of diaminopropane and putrescine, whereas some ion-pairing methods do not (Wagner et al. 1982; Seiler 1983b). Ion-pair reverse phase HPLC with UV monitoring of underivatized compounds followed by detection of post-column OPA derivatives is a very sensitive method. Detection of underivatized SAM metabolites by UV adsorption and polyamines as OPA derivatives by fluorometry is achieved at the picomole level (Greenberg and Cohen 1985).

5 Ion-Exchange Chromatography of Underivatized Polyamines

Polyamines in low pH media are strong cations, with net positive charges, and can be separated on a cation-exchange column based on differences in their electrostatic affinities for the stationary phase. A wide range of amine compounds in plant tissue can be separated and quantified by ion-exchange chromatography, coupled with post-column derivatization with o-phthalaldehyde. This approach has been employed for separating polyamines, aromatic and heterocyclic amines, polyamine homologs, and other related compounds in biological tissues (Villanueva and Adlakha 1978; Aleksijevic et al. 1979; Adlakha and Villanueva 1980; Perdrizet and Prevost 1981; Tabor and Tabor 1983; Villanueva et al. 1977, 1978, 1985, 1986a, b). Thus ion-exchange chromatography is potentially useful for studies which follow aliphatic polyamines and aromatic amines in cinnamic acid and alkaloid biosynthesis in plants.

5.1 Chromatographic Separation of Underivatized Amines and Related Compounds

Crude extracts of plant tissue can be prepared using procedures similar to those described earlier and analysis can be performed on samples without pre-purification. Resolution of a large number of substances in a single sample is achieved on an amino acid analyzer with a cation exchange column, by eluting with a two-step or three-step buffer system, with the appropriate pH, ionic strength, and organic solvent, using also a two-stage temperature regime (Fig. 7) (Villanueva and Adlakha 1978). Chromatographic separations may require 2 to 3 h in addition to 40 min for column regeneration and re-equilibration. In an improved high performance chromatographic method, amines and related compounds are separated in a much shorter time, 72 min, on a 9×0.45 cm i.d. column of BCX-12 cat-